Life Before Man

LIFE WORLD LIBRARY
LIFE NATURE LIBRARY
TIME READING PROGRAM
THE LIFE HISTORY OF THE UNITED STATES
LIFE SCIENCE LIBRARY
GREAT AGES OF MAN
TIME-LIFE LIBRARY OF ART
TIME-LIFE LIBRARY OF AMERICA
FOODS OF THE WORLD
THIS FABULOUS CENTURY
LIFE LIBRARY OF PHOTOGRAPHY
THE TIME-LIFE ENCYCLOPEDIA OF GARDENING
THE AMERICAN WILDERNESS
THE EMERGENCE OF MAN
FAMILY LIBRARY
THE TIME-LIFE BOOK OF FAMILY FINANCE
THE TIME-LIFE FAMILY LEGAL GUIDE

TIME LIFE BOOKS ®

The Emergence of Man

Life Before Man

by the Editors
of Time-Life Books

Time-Life Books
New York

THE EMERGENCE OF MAN

SERIES EDITOR: Carlotta Kerwin
Editorial Staff for *Life Before Man*:
Text Editors: Jay Brennan, William Frankel,
L. Robert Tschirky
Picture Editor: Jane Scholl
Designer: William Rose
Staff Writers: Paula Pierce, Suzanne Seixas,
Timberlake Wertenbaker
Chief Researcher: Peggy Bushong
Researchers: Edward Albert, Muriel B. Clarke,
Susan Jonas, David Palmer, Wendy A. Rieder
Design Assistant: Lee Wilfert

Editorial Production
Production Editor: Douglas B. Graham
Quality Director: Robert L. Young
Assistant: James J. Cox
Copy Staff: Rosalind Stubenberg,
Susan B. Galloway, Florence Keith
Picture Department: Dolores A. Littles,
Marianne Dowell

Valuable assistance was given by the following departments and individuals of Time Inc.: Editorial Production, Norman Airey, Nicholas Costino Jr.; Library, Peter Draz; Picture Collection, Doris O'Neil; Photographic Laboratory, George Karas; TIME-LIFE News Service, Murray J. Gart; Correspondents Margot Hapgood (London), Ann Natanson (Rome), Maria Vincenza Aloisi (Paris), Edmon Ogle (Sydney).

The Authors: PETER WOOD (Chapter 1) is a writer on the staff of TIME-LIFE BOOKS.
LOUIS VACZEK (Chapter 2) is the Senior Science Editor of the *Encyclopaedia Britannica*.
DORA JANE HAMBLIN (Chapter 3), a freelance writer, was a staff writer for LIFE magazine for many years.
JONATHAN NORTON LEONARD (Chapters 4 and 5), a freelance writer, was Science Editor of TIME magazine and is author of a number of volumes published by TIME-LIFE BOOKS, including *The Planets, Ancient America* and *Atlantic Beaches*.

The Consultants: A. W. Crompton, chief consultant, is Director of the Museum of Comparative Zoology, Harvard University. Farish A. Jenkins Jr. is Associate Curator of Vertebrate Paleontology and Robert T. Bakker is Teaching Fellow at the same museum. Theodore Delevoryas is Professor of Biology in the Osborn Memorial Laboratories, John H. Ostrom is Professor of Geology and Geophysics in the Kline Geology Laboratory, and Elwyn L. Simons is Professor of Geology and Geophysics in the Peabody Museum of Natural History, all at Yale University.

The Cover: Tyrannosaurus rex, or king of the tyrant reptiles, which ruled the earth some 75 million years ago, peers from a ferny thicket. The scene has been re-created by superimposing a painting of the giant predator on a photograph of a type of forest known to have existed in the Age of Dinosaurs. The same technique has been used on pages 148-149.

Published simultaneously in Canada.
Library of Congress catalogue card number 72-86602.

Contents

Introduction

A man is a man, and a woman a woman, but both are also primates, mammals, vertebrates, chordates and metazoans. Those are not merely words invented by professors to beguile the public. They are a way of saying that man, whatever additional he may be, is an animal. He has something in common with all animals. The microbes in the ditch, the ants on the lawn, are relatives of man, and his development cannot be fully understood without reference to them. How he rose from a rare, not especially impressive animal to his present status of dominance is the subject of The Emergence of Man, the series of books of which this book is a part.

The story is a tortured, twisting one. It must account for the strange fact that man, and not the lordly dinosaurs that ruled when man's furry ancestors first scurried about, survived to command the earth. And before the tale ends with modern man—living on a planet that he has already modified, not always for the better, and has the power to destroy—it must trace the origins of ideals and rituals, prayer and cannibalism, tools and war, gods and empire, trade and farming, and all the facets of life that make man human. But it begins with his biological background, treated in this volume.

The concept of human evolution is as old as Charles Darwin's *On the Origin of Species,* which was published in 1859—even older in a timid way. But today knowledge about the mechanisms of evolution is accumulating at an unprecedented and ever-increasing rate. Part of the new understanding comes from new scientific techniques. With an electron microscope, for instance, a virus so small that its existence had to be indirectly inferred is made to show up as clearly as the windows of a building across the street. Each such new view reveals more clearly how man came to arise from the simplest forms of life in the primeval waters.

Great advances have also been made in dating methods. Until well into the present century most figures given for the age of fossils or the remains of ancient man were hardly better than guesses. But recently estimates have been replaced by accurate measurements. One method, which uses the radioactive decay of carbon, provides a reliable time scale nearly 40,000 years into the past for objects containing carbon, such as campfire charcoal. For dates ranging into millions of years, a method known as potassium-argon dating is being increasingly used as a guide to the age of rocks and many objects, such as bones, that may be embedded in them.

One of the most fruitful of the new ways to learn about human antecedents is to observe—as this book does—living animals that resemble man's direct ancestors. Among these distant cousins of man are tree shrews: primitive animals not very different from the earliest mammals. Another is the coelacanth, a rare fish descended from ancestors that had inside their fleshy fins bone connections uncannily like the bones of human arms and legs. On limbs much like these the first vertebrates crawled up on the land.

Today even animals distant from man can reveal insights into his past. In particular, much about ancient behavior is deduced from studies of modern animal behavior. Man is a social animal, for example. He was not, however, the first to find strength in numbers. Several types of insects did so many million years ago, and the result was the wonderful world of the social insects—ants, bees, wasps and termites—whose "civilized" colonies can be found in every inhabitable part of the earth. Though the insects provided none of man's heritage, their group living offers illuminating parallels to his own societies.

Similar parallels can be found in the tightly structured group living of such animals as wolves and baboons. But none of these low-level societies of mammals, interesting as they are, show signs of progressing to a higher level. This feat, which literally changed the face of the earth, was accomplished by smallish, erect-walking primates who were the direct ancestors of man. Their hunting groups, which at first were presumably as simple as wolf packs, gradually became more tightly organized. Their descendants developed speech for quick and accurate communication. They learned how to use fire and fashion weapons of wood, stone or bone. They built shelters to protect themselves from inclement weather and acquired clothing that enabled them to live comfortably in cold climates.

From this point onward, the history of man is largely that of his technical advances and social achievements. Perhaps the greatest achievement was the almost simultaneous development of agriculture and animal husbandry. When the first farmers had acquired domesticated plants and animals, they turned unproductive land into cultivated fields and pastures. Human population increased enormously and pushed into areas inhabited thinly by wandering hunters. Villages appeared, grew bigger, acquired walls for protection and temples for local gods. Then came cities; then empires. In not much more than 1.3 million years—a short time on the evolutionary scale—from the appearance of the first creature that could be called human, man had changed from a scarce and wandering hunter to undisputed lord of his planet.

—The Editors

Chapter One: The Paragon of Animals

After 3.5 billion years, Homo sapiens sapiens, thinking man, emerges before the energy-giving sun to become the dominant species.

The house lights have dimmed. The stage is a black void. The rustling of programs and the murmur of gossip subsides. Silence. Gradually a figure appears; ghostly, transparent at first, then more and more substantial, solid, radiant at last, shining out of the darkness. It is man—the hero of this story.

Shakespeare glorifies him as only Shakespeare could: "What a piece of work is man! How noble in reason! how infinite in faculty! in form, in moving how express and admirable! in action how like an angel! in apprehension how like a god! the beauty of the world! the paragon of animals!" Yet in his very next words the poet could not resist asking the question that all of us, at one time or another, have asked ourselves, "And yet, to me, what is this quintessence of dust?"

The question is as old as man and has been answered in nearly as many ways as there have been men to pose it. In the technical jargon of biological classification, modern man is *Homo sapiens sapiens* —a Latin label that means only "intelligent man." More informatively he has been called a political animal, a tool-using animal, a social animal, a creature that is aware of itself—and these are but a few of the aphorisms with which men through the ages have sought to nail down what it is to be human. Men are all these things, of course—and more. From a purely materialistic point of view, for instance, a man—any man—represents the most complex assemblage of molecules ever to appear on earth, possibly in the universe. In this respect an individual man differs from other organisms only in degree. But collective man—that is, man organized in social groups—represents a quantum leap beyond all other organisms. For the moment, at least, he is in command of the spaceship Earth, perhaps in danger of wrecking it even before liftoff, but equally able to steer for the stars.

To understand how man came to seize the controls is to answer Shakespeare's question. The story is complex, full of surprising twists, and strangely long, beginning at the moment when life first appeared on earth more than 3.4 billion years before man himself existed. This last third of the 20th Century is an especially good moment to trace the story, for this era is seeing a new phase in the study of man. In the past the most meaningful descriptions of the human state were made by prophets, artists, philosophers and poets. Theirs were personal views, colored by personal, subjective biases. We do not lack for such descriptions today, but at the same time we are gaining another perspective upon man, an objective view seen through the lens of modern science. The lens does not present a fixed image; instead, it builds an expanding mosaic of exquisite details, less poetic, perhaps, but no less awe-inspiring than Shakespeare's description.

New fossil evidence of man's ancestors, for example, is turning up at an unprecedented rate in regions such as eastern Africa, allowing us to trace the steps by which humans arose from less-than-human forebears. In 1859, when Charles Darwin propounded the landmark theory that underlies our present understanding of man's evolution, scientists knew of exactly two fossils that were relevant to the search for man's origins: one of an extinct ape, another of the early type of Homo sapiens called Neanderthal man. Hardly more than a century after Darwin's book appeared, expeditions in the Lake Rudolf area of East Africa unearthed more than 150 near-human bones in a single five-year period. One of these bones, the so-called Lothagam jaw, is 5.5 million years old, a date that pushes the record of man's certain ancestry more than a million years further back than any previous find.

The paleontologists' hunt for fossils has been aided by knowledge and insights drawn from other sciences. Atomic physicists, studying the rates of radioactive decay in various natural substances, have given paleontologists new and more accurate methods for pinpointing the stages in the evolution of life. Scientists can now determine the age of volcanic rocks by measuring the transformation of radioactive potassium into argon gas within the rock; the amount of argon found indicates how much time has passed since the rock formed and its potassium started to change. In a similar way, somewhat younger materials that once were alive, such as wood and bone, can be dated by measuring the transmutation of a radioactive form of carbon to another substance.

Equally valuable have been the contributions of modern biochemistry. Not until 1966 did biochemists finally decipher the genetic code—the complex structure and functions of a substance called DNA, which is present in virtually all living organisms. Through DNA, instructions for the building of new cells and new organisms are formulated and passed along. And having cracked DNA's code, scientists can begin at last to understand two contrasting mechanisms of evolution: mutation, in which minute variations in DNA instructions may originate new species of animals and plants; and genetic invariance, the precisely accurate duplication of DNA instructions without variation, generation after generation, that enables members of existing species to reproduce themselves essentially unchanged.

At the very frontiers of modern biochemistry, DNA is yielding new secrets to researchers. One of the most exciting of them is the process by which, over millions of years, mutations gradually create subtle differences in the structures of proteins, the basic building materials of all living things. Some scientists believe these differences accumulate at a steady rate and thus can be used to measure the evolutionary separation between man and other species. In the blood substance called hemoglobin, for example, the proteins of a horse exhibit no fewer than 42 differences from those of a man; clearly, the ancestors of man and horse parted company as distinct species a long, long time ago. By contrast, the hemoglobin proteins of man and monkey exhibit only 12 differences, while those of man and chimpanzee have none at all. Obviously, man is close to the apes, less close to monkeys, still less close to horses. But scientists already knew that.

What is exciting in the new knowledge is the possibility of working out a sort of protein clock that would indicate the time at which all existing species of animals first emerged. Though the protein clock is still tentative and experimental, it offers the hope of a dating method supplementary to the older techniques that depend on fossils, radioactivity or the differences between layers of rock.

Other clues to the past are coming from studies of a very different kind involving a host of living animals. The science of animal behavior is a relatively new discipline, but it is a flourishing one, and its basic materials are peculiarly accessible to the layman. Consider, for example, Jane van Lawick-Goodall's description of the greeting rituals among chimpanzees on a reservation near Lake Tanganyika, in Africa: "When two chimpanzees greet each other after a separation, their behavior often looks amazingly like that shown by two humans in the same context. Chimpanzees may bow or crouch to the ground, hold hands, kiss, embrace, touch, or pat each other on almost any part of the body. . . . A male may chuck a female or an infant under the chin. Humans, in many cultures, show one or more of these gestures."

Observations like Jane van Lawick-Goodall's help explain the basis for some human behavior, particularly in social actions, and also suggest how ancestral man may have acted, and why. Studies further from the human family tree are no less significant. Even insects tell something of how life can be organized. And wolves, like man, have evolved complicated life styles based on the cooperative hunting of game and the sharing of the kill. As individuals wolves shed little light on humans, but their hunting strategies, their hierarchical social structure, their divisions of labor and their territorial jealousies help explain similar patterns in early man.

From studies like these, a new view of man and of man's ancestry has been emerging. It places man in perspective in a vast span of millennia amid a vast crowd of creatures, and it shows something of why he is, as Shakespeare said, the "paragon among animals." But before we turn to distant places and distant times, before we bring on stage the cast of millions, let us look at the finished product, the hero of the epic, isolated on a dark stage. For the moment our concern must be more limited than Shakespeare's. We cannot completely answer "What is man?" until we can answer a simpler question: "What makes man different from other creatures?"

His mind, to be sure. But what our new knowledge makes ever clearer is that the mind is not enough. Without a remarkable combination of organic hardware that supports and abets it, the mind would be useless. Man dominates the animal kingdom not only because he is blessed with a big brain but because of a special combination of physical characteristics that is often taken for granted. Beside the sleek grace of a jungle cat, the streamlined strength of an 1,800-pound tuna or the regal bearing of a horse, what is man's puny body? The answer to that rhetorical question, as a careful examination of man's physical adaptations will illustrate, is: everything.

Among the physical traits that together separate all men on the one hand from all other animals on the other, there are three of overwhelming significance: a skeleton built for walking upright; eyes capable of sharp, three-dimensional vision in color; and hands that provide both a powerful grip and nimble manipulations. Controlling and making use of this equipment is the brain—a physical organ itself, but one that introduces the capacity for rational thought and, with the body, makes possible that other most human of all man's distinctive abilities, speech.

These attributes, uniquely combined in man, interact with one another. It is impossible to say that one led to the next, or that one is necessarily more important than the others. They developed together, each reinforcing the others and making possible improvements in them. Nevertheless, one attribute stands out simply because it is so conspicuous: upright walking. It is a remarkably effective method of locomotion, and no animal can use it as man does.

Man versus Runners, Jumpers, Walkers

The four animals at right are all equipped for effective locomotion over the level land they normally inhabit. Yet each moves in a quite different way. The long-legged ostrich can run at a speed of 50 miles an hour; the kangaroo can hop in 40-foot bounds; the pig's rocking gait, which he shares with other quadrupedal mammals, covers long distances with little effort.

Man, upright on two legs, cannot match the specialized gaits of ostrich, kangaroo and pig. But his unique anatomy enables him to make use of all three advantageously: He can run at 15 miles an hour for several minutes and can attain a maximum speed of more than 20 miles an hour over short distances; he can broad-jump 29 feet and walk 50 miles or more in a day—to say nothing of swimming rivers and climbing mountains. In addition, his upright movement frees his hands for tasks that give locomotion another dimension of usefulness.

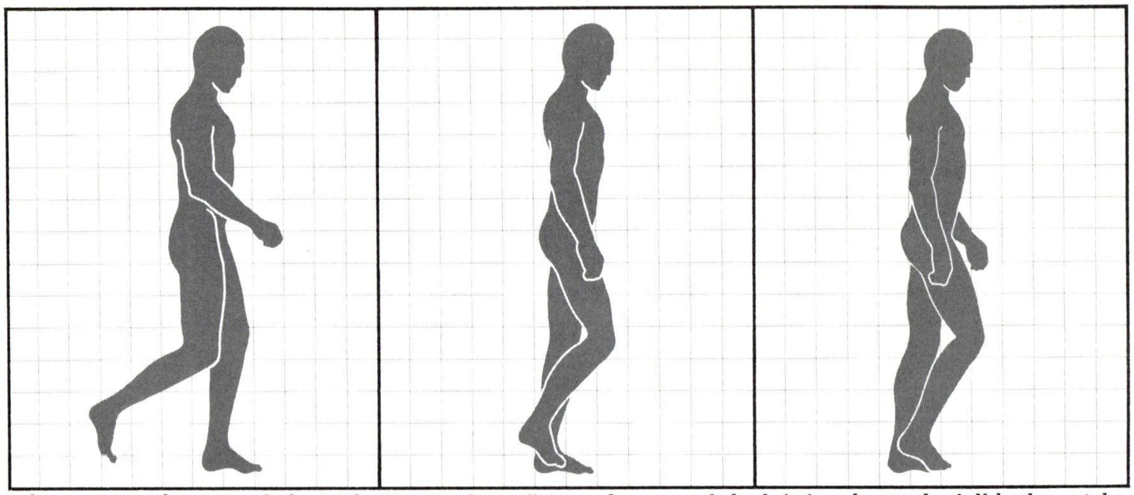

When a man takes a stride his right foot pushes off from the toe and the left foot bears the full body weight

Ostriches step out using alternate legs and balance on big feet, as men do, but they stand on their toes, and

The kangaroo crouches, straightens both powerful hind legs—like a mousetrap snapping—to take off, then

Lifting one leg at a time when walking—right hind foot, right fore foot, left hind foot, left fore foot—the pig

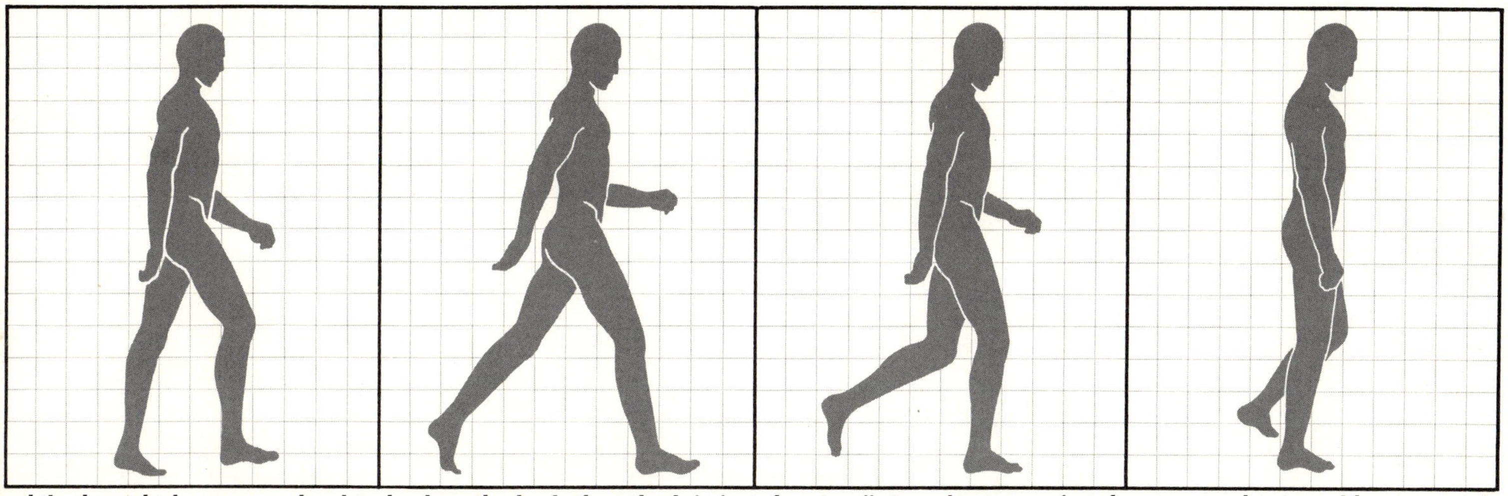

while the right leg moves ahead to land on the heel; then the left foot thrusts off. In order to run fast, he stays on his toes, like an ostrich.

their ankles are high above the ground; the stepping motion thus employs a very long "lever," which makes for a long stride—and high speed.

swings the legs up and ahead while it soars in a long two-legged jump. A man making a standing broad jump uses a somewhat similar action.

rocks from side to side to avoid falling, keeping its body weight over the tripod defined by three legs while the fourth is commencing a step.

Let us begin, then, with the act of walking. For all its apparent simplicity, it is an adaptation as specialized as flying is to a bat or swimming to a seal. True, man is not the only animal able to stand on its hind legs alone; birds, bears and a number of man's primate cousins are bipedal on occasion. But with the exception of a few flightless birds, such as the ostrich, man is the only animal that depends exclusively on two legs for locomotion—whether crossing a room or crossing a continent, moving at high speeds or aimlessly strolling, with arms burdened or swinging free. Using his two legs, a man has the endurance to outrun a deer. He can carry heavier loads, pound for pound of body weight, than a donkey—the French-Canadian *voyageurs* who transported Indian trade goods through the North Woods routinely back-packed 180 pounds of bales over nine-mile portages, and a legendary hero among them named La Bonga is said to have portaged 450 pounds. No terrain is totally impassable to a man; he can reach an eagle's aerie or a pearl oyster's bed. Only a man, the British scientist J. B. S. Haldane noted, can swim a mile, walk 20 miles and then climb a tree. When compared with the versatile and powerful scheme of human locomotion, even the regal movement of a horse turns out to be limited indeed.

Like horses, men have a variety of gaits; they amble, stride, jog and sprint. Among them all, though, the simple stride is at once the most useful and the most peculiarly human way of getting from one place to another. Developed on the African savanna, where man's early ancestors often covered many miles in the course of a day's hunting, the stride has taken man to every corner of the earth. It is no minor accomplishment. When compared with the way four-legged animals get about, human walking turns out to be a surprisingly complex feat of acrobatics. "Without split-second timing," says John Napier, a British authority on primates, "man would fall flat on his face; in fact with each step he takes, he teeters on the edge of catastrophe." Human walking is actually a balancing act in which the muscles of the feet, legs, hips and back are alternately contracted and relaxed according to synchronized orders from the brain and spinal cord.

It is all uniquely human, this "heel-and-toe-and-away-we-go" cycle, and to those who can see it with fresh eyes, it is strangely beautiful. Uniquely human, because no other creature on earth can do it. And beautiful in its sheer efficiency, in its superb adaptation of bone and muscle, brain and nerve, to the tricky problem of moving about on two legs rather than four. The adaptation was achieved at considerable cost. Back trouble, for one thing, is common among men, and comes partly from upright posture.

But why is it so important to man that he stand erect and walk on two legs? Part of the answer has to do with man's head, another with his hands. The advantages for the head are often overlooked, yet simply raising the head high above the ground has had crucial results. The head is where the eyes are, and the taller a man stands the more he sees. A dog running through tall grass is forced to leap into the air time and again to find his bearings, but even on a smooth surface, where no obstacles obstruct vision, the advantage of height is enormous. Eyes that are two feet above ground level can detect objects two miles away; eyes five feet above the ground can see a mile farther.

The advantage of height is especially important because vision is by far the most directly useful of man's five major senses.

Scientists estimate that some 90 per cent of all the information stored in the brain arrived there through the agency of the eyes. Not surprisingly, man's eyes are attuned precisely to his needs. For general seeing they are unsurpassed by any in the world. A hawk may see more sharply but cannot move its eyes easily and generally moves its head to follow its prey. A dragonfly can follow faster movement than a man but cannot focus a sharp image. A horse can see almost completely behind its head but has difficulty seeing objects straight ahead at close range. Most important, among higher animals only man and his nearest primate relatives have the special combination of full stereoscopic and color vision. Man's eyes, placed at the front of his head rather than the sides, can focus together on an object so that it is perceived as a single three-dimensional image in the brain. Within this image his color vision enables him to pick out details by hue as well as by form and brightness.

Taken together, color and depth perception bring man enormous advantages over most other animals, the majority of which are color-blind and have a relatively poor capacity to judge visual distances or focus in fine detail upon particular objects. What a hunting dog sees when it looks out over an open field is little more than what a black-and-white movie might show and his distance focus is limited. If there is a rabbit in the field, the dcg is unlikely to spot it unless it moves—one reason why rabbits and similar prey freeze to conceal themselves from their enemies. A human hunter, on the other hand, can scan a scene from his feet to the horizon in a few seconds by focusing sharply and selectively upon a succession of different images. And he sees more images than any dog does because his eyes are raised at least three feet higher above the ground.

But if man stands up partly in order to see, and stays up partly because he sees so well, the freedom that his posture gives to his arms has proved even more decisive. Chimpanzees, among man's closest competitors in upright posture and bipedal movement, have never really mastered the art of walking on their hind legs, and they lack man's free use of the arms. For a brief while they can get about in their forest homes with a bunch of bananas or an infant in their arms, but they must always be ready to assist their balance with the help of a knuckle on the ground. On the other hand, man, who learned very early how to walk in open country, has thrown caution to the winds. Babies may crawl on all fours; old people may rely on canes; but most men go about with never a thought of support from anything but two legs: their hands are free to grab and use things.

The hand that is not needed for support can take on more responsible and more creative tasks, and it has become the instrument by which man has prospered. With 25 joints and 58 distinctly different motions, it represents one of the most advanced mechanisms ever produced by nature. Imagine a single tool that can meet the demands of so many different tasks: to grip a stick, to play a concerto, to wring out a towel, to hold a pencil, to gesture and —something we tend to forget—to feel. For, in addition to the ability to perform tasks, the hand is our prime organ of touch. In the dark or around corners, it substitutes for sight. In a way, the hand has an advantage over the eye, because it is a sensory and a manipulative organ combined into one. It can

Four Views of the World

The photographs at right show how differently a man, a dog, a horse and a bee see the same sunlit grove. (The horizontal visual field of each has been assembled by lining up photographs specially made to represent animal vision).

Man sees the smallest part of the grove—but in that section he sees the most. The human visual system distinguishes among some 10 million gradations of color; it also can adjust to the 10-billion-fold range between the dimmest thing it can discern and the brightest object it can see without pain; it focuses to see sharply either the nearest ferns or the most distant tree. And man's vision has one more quality the pictures here cannot show: the sense of depth provided through his broad, stereoscopic field of view.

A spring scene appears to man's 180° vision in full color, the green

The dog's color-blind eyes perceive a broader and in one way sharper field—nearby leaves are

A horse's wide-set eyes cover everything in a 360° view that is interrupted only by a narrow area directly behind the head. Though the horse does

A honeybee sees a fuzzy pattern made by the thousands of lens-tipped cones of its compound eyes. Its vision spans a circle—minus parts blocked

leaves sharp at center but blurred on the trees at far left and far right.

Dogs, descended from twilight hunters that relied mainly on scent, see only shades of gray, and even these are indistinct at a distance.

The horse's wrap-around field of vision alerts it to enemies in every direction; its whole world is in sharp focus from a distance of four or five feet to the horizon—but it is colorless.

The bee makes do with a different kind of vision, using compound eyes comprising thousands of tiny immobile lenses clustered on its head. Each lens receives a pinpoint view of whatever is in its direct line of sight, and the combined impressions of these lenses make up a blurry pattern. A bee also sees only certain colors: yellow and green (seen as one color), blue-green, blue, violet and some ultraviolet shades that man is unable to detect.

distinct from far left to far right. But the dog is shortsighted, and background trees are blurred.

not perceive color, it distinguishes fine shadings of gray such as those between the open clearing (left) and the dense undergrowth (center).

by its body (far left and far right)—but is sharp only close up (sixth picture from left), where hundreds of lenses bear on the same object.

explore the environment by means of touch, and then immediately do something about what it detects. It can, for instance, feel around on a forest floor for nuts and roots, seize them on contact and pop them into the mouth; at the same time that your eyes are reading these words, your hand can finger the corner of the page in preparation for turning it.

A marvelous tool the hand itself may be, yet its full value is exploited only when it is employed to manipulate still other tools. This capacity is a second-stage benefit of upright walking. With an erect posture, man's hands are free; with hands free, he can use tools; with tools he can defend himself better and get food more easily. Humans are not the only animals that employ tools, but they are the only ones that have two distinct ways of holding and using them: the "power grip" and the "precision grip" in John Napier's terminology. In the power grip an object is held between the undersurface of the fingers and the palm of the hand. In the precision grip it is held between the tips of the fingers and the opposed thumb. Infants and children begin with the power grip and progress to the precision grip. Think of how a child holds a spoon: first in its fist (the power grip), and later between the tips of the thumb and first two fingers (the precision grip). It is significant, perhaps, that some civilized peoples place great importance on the way in which their children hold their spoons, as if the grip were symbolic of the difference between having an ape at the table or having one of their own maturing offspring.

Many primates, in fact, share the power grip with man. It is the way to get firm hold of a tree branch. But neither a monkey nor an ape has a thumb long enough or flexible enough to be completely opposable, able to reach comfortably to the tips of all the other fingers, as is required for the delicate yet strong precision grip. It is the human thumb that makes possible nearly all the movements necessary to handle tools, to make clothing, to write with a pencil, to carry a suitcase, to play a flute.

If the precision grip required to play a flute can be related to upright walking, then the mind required to make such music may be related to the grip. For tools and brain seem to have developed together. It is the hand that carries out some of the most critical and complex orders of the brain, and as the hand grew more skillful so did the brain.

The human brain is not much to look at. On the dissecting table, with the skull removed, it is, in the words of one observer, a "pinkish-gray mass, moist and rubbery to the touch . . . perched like a flower on top of a slender stalk." (The stalk is the spinal cord, which may be considered an extension of the brain.) In appearance, an ape's brain does not differ too greatly. But there is a difference, and it is crucial. It is in the gray layer, called the cortex, the outer layer of the largest part of the brain. The cortex, scientists now know, plays the major role in reasoned behavior, memory and abstract thought—and also supervises the delicate and accurate muscular movements that control the precision grip. The cortex is quite thin, but it represents 80 per cent of the volume of the human brain; if spread out flat, it would be about the size of a newspaper page. It fits inside the head only by being compressed like a crumpled rag —the famous "convolutions" of the brain are in fact mainly the folds and overlaps of the cerebral cortex —and its compression bespeaks the fact that it has all but outgrown its allotted space. Somehow, this increase in the size of the cortex has made man's brain the uniquely human thing it is. Says anthropologist William Howells: "We . . . do not know in what way a larger brain makes us more intelligent. But it has clearly done so." If there are many mysteries about the brain that remain to be solved—and there are, in plenty—the main secrets and the main importance of man's huge cortex are now well understood.

The cortex is not only the seat of intelligence; it is also, and perhaps more significantly, the association center of man's brain. That is, it is the part of the brain where sense impressions and memories are stored—to be called forth and acted upon as circumstances suggest. There is no fixed pattern in which these associations need be made, as there is in animal brains, and no predetermined result of their calling forth. Among animals, many patterns of action are nearly automatic, performed by instinct or through previous conditioning. In man, for the most part, these patterns are performed consciously, or refrained from consciously, or replaced by completely new patterns, again consciously. This use of the brain results in what is known as reasoned behavior, a mental phenomenon that only man is capable of because only he has the large cortex that is necessary in order to achieve it.

The great brain gap between man and other animals can be visualized by looking at what happens to both parties when a man's hand pokes the outspread tentacles of a sea anemone. The anemone will instantly retract its tentacles into its body; the reaction is automatic, since what passes for a brain in the anemone is programmed for only one pattern of action: in response to touch the tentacle retracts. No reasoned behavior is involved. In response to the same contact, the man may pull his hand back, or he may not. His brain considers options, and his action will depend on many things—whether he thinks anemones are dangerous or harmless, whether the contact is pleasing or discomforting, whether he touched the anemone on purpose or accidentally. Most higher animals can also react to a given stimulus in a variety of ways—but not a single animal has anything like the number or diversity of man's potential responses. And man is completely alone in his capacity to examine all of his options in advance, to look inward upon himself and to observe the processes of his own mind—in short, to think. Perhaps even more important, when he thinks, he knows he is thinking.

While conscious thinking is one of the proudest badges of human superiority, it remains one of the most puzzling. We cannot yet explain the operation of the brain cells the way we can analyze the movement of bones and muscles in walking and grasping. But a start has been made. Thinking depends on association and memory in the cortex. Ideas and thoughts are registered in the nerve cells, or neurons, somewhat as they are in a man-made computer, in the form of electrical patterns, and they are retrieved and shuffled about by electrical actions. This much is quite clearly established, since thinking produces measurable electric currents in the brain and many experiments demonstrate the effect of electric stimulation in such processes as memory. Electroshock therapy of a schizophrenia victim, for example, can erase some of the patient's recent memory while leaving unaffected recollections that date from the more distant past. The brain evidently has two memories, as a computer does, one for storing considerable in-

A Handy Way for the Inarticulate Chimps to Talk

Only man can actually speak. But though chimpanzees, man's closest relatives, cannot learn to talk, they can be taught to communicate with man by using the American Sign Language. In the examples shown here, chimps use the symbols (translated in drawings adapted from the American Sign Language) mostly to express the desire for food, affection or attention. But they are also able to deal with such other matters as "key," "tree" and "hat."

In the "tree" sign, one hand holds the opposite forearm upright by the elbow, and the free hand is fluttered back and forth.

The sign meaning "hat" is made by first placing the hand on top of the head and then making a repeated patting motion.

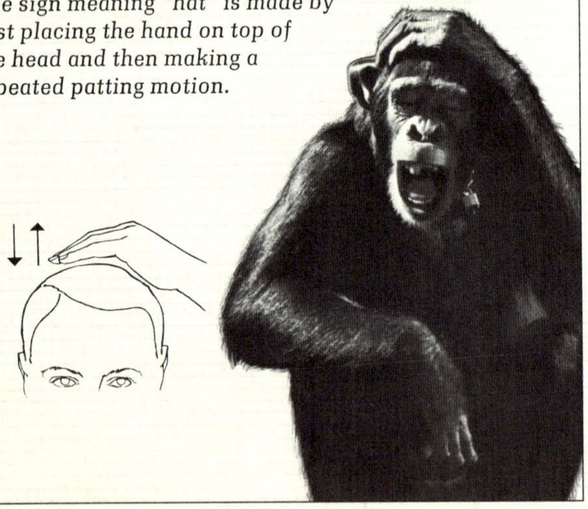

To make the gesture that stands for "fruit," the fingers are loosely curled and the hand is drawn down the side of the cheek.

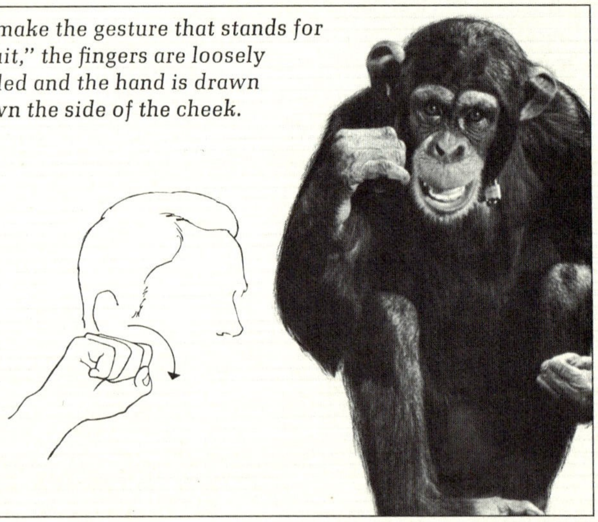

formation more or less permanently, another for temporarily recording current data.

The brain's similarities to a computer are remarkable, but they are only coarse similarities; comparing a brain with a computer is much like comparing an aircraft carrier with a bark canoe. The human brain contains an estimated 10 billion nerve cells, and each of these cells may be thought of as a switching point for the electrochemical signals of mental activity. The largest modern computer, by contrast, contains 1.5 million switching points. The circuitry within the brain is obviously many thousands of times larger and more complex than that of the most complicated computer yet devised. As Warren McCulloch, an American student of the brain, has put it, "The brain is like a computing machine, but there is no computing machine like the brain."

If the brain is more than a computing machine, it is also more than a thinking machine. Reasoned behavior itself did not make man the paragon of animals. He rose to dominance as his extraordinary brain interacted with his superior body to make possible crucial physical achievements. The ceiling of the Sistine Chapel was painted by a precision grip and color-sensitive eyes controlled by Michelangelo's brain. Neither his bodily machinery nor his creativ-

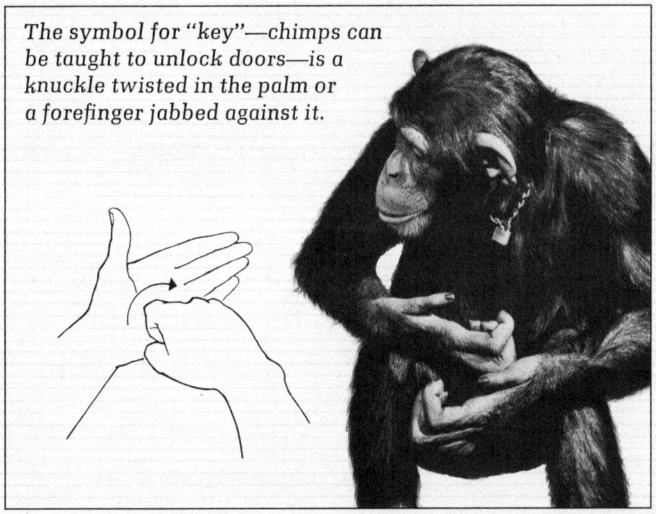

The symbol for "key"—chimps can be taught to unlock doors—is a knuckle twisted in the palm or a forefinger jabbed against it.

By pointing to the corner of its eye, a chimpanzee can signal "see." The same gesture is also used to signify "look."

The sign symbolizing "touch" has been adapted for chimps, which will stroke the back of a hand when they want to be tickled.

A chimp puts its forefinger on its ear to make the sign that translates either "hear" or "listen."

ity alone could have produced this masterpiece; both were needed, working together.

The great significance of this combination of human brain and human body is perhaps best shown by man's most important innovation: language. Only humans can talk, although all animals communicate with their fellows. Bees dance to direct the swarm to food; wolves warn off intruders by marking their territories with scent; one bird call announces danger, another invites love-making. (Chimpanzees can be taught to use human sign language.) Besides employing all of these primitive methods of communication —odors, bodily movements, simple sounds—humans also use language, a huge repertory of sounds that can be combined as units to express very complex facts and ideas. The prairie dog's quick, high-pitched barks can send up a vague alarm *(page 142)*; they cannot specify: "Five men armed with shotguns are approaching from the west and will be upon us in half an hour."

Such communications obviously depend on the brain, for some lower animals equal man in vocal performance without mastering language. Myna birds and parrots can mimic a man's voice perfectly; they can even be taught to repeat sentences of several words or more—but they cannot really talk, because their brains are incapable of abstract thought. They cannot, therefore, combine elements from two different sentences, learned by rote, and use these elements to construct a third sentence.

Language is so clearly dependent on brainpower that its equal dependence on the body is often overlooked. The role of the body is most clearly demonstrated in the case of chimpanzees. They seem to have brains that are adequate for abstract thought. A chimp, for example, can stack several boxes on top of one another to reach a bunch of bananas, a simple act requiring the imaginative combination of superficially unrelated elements. It also can produce a wide range of sounds. It ought to be able to talk. Since the turn of the century scientists have been trying to teach chimpanzees to speak. The best anyone has been able to do, after years of patient tutelage, is to get a chimp to say "mama," "papa" and one or two other infant words. Only recently has the reason for this failure been traced; it is not simply brain size but another aspect of the anatomy. A close examination of the chimpanzee showed that it lacked, among other things, the pharynx that enables humans to articulate vowel sounds. The work of scientists at the University of Nevada has shown that chimpanzees are indeed able to construct simple sentences —but not spoken sentences. They can "speak" with visual rather than auditory symbols—the symbols of the American Sign Language, originally designed for the deaf. Man remains the only creature that has developed both the physical structures and the powerful, specialized brain needed to produce speech.

Learning to talk was the last of man's major evolutionary achievements. And with the gift of speech, man acquired an immensely powerful tool for speeding up his cultural evolution. The foundations of human civilization could be laid. From the beginning, the members of man's hunting and gathering bands made good use of their ability to communicate verbally with one another—to plan a hunt, pass on information or agree upon a rendezvous. But the greatest benefit man gained through language came later, through the ability to learn from the accumulated experience of others—other people and other groups. Before the birth of language, man's experience was pitifully brief and transitory; when a man died, his experience died with him. By the gift of language the shared experience of mankind could be preserved and kept accessible over the course of many generations, first through recited lore and legends, later through the written word.

How important this gift of language has proved can be seen in a quick glance around. Physically, modern men are hardly distinguishable from men who lived 30,000 years ago. But socially, human life has been transformed by the accumulation of the experience of millions upon millions of human lives over thousands upon thousands of years. This new world is all based on words. From a species surviving in a tropical savanna, man has come to occupy the entire globe. From an estimated population of 10 million as recently as 10,000 years ago, mankind has multiplied to 3.6 billion today, and threatens by its very success to exhaust the resources of the earth before the next century is well advanced.

This is man as he stands today, unique among the animals and alone in command of his planet. The foot that evolved from a branch-gripping prehensile appendage to a limb capable of carrying a man steady and erect over a rolling grassland now, encased in a boot, takes him slogging through freezing city slush. The hand that first wielded a stick as a weapon and later chipped flint into a cutting edge today fashions tools that make tools that make more tools that make rocketships that reach the planets. The eye that used to spot a wounded giraffe, hiding in a copse across a plain, now scans this page. And the same mind that learned to analyze the migrations of game, to recognize dozens of different animal spoors, to distinguish among hundreds of varieties of plants, now dictates the playing of a game of chess, the writing of a book, the waging of a war.

It seems an impossible journey from the African savanna to Cape Kennedy, from the first stone tools to a room-sized computer. Yet it was completed in less than three million years—an instant in the long history of life.

Perhaps the most fruitful approach to the problem is to retrace the steps by which body and brain evolved. We human beings are primates, sharing many characteristics with the monkeys and the apes. We are mammals, warm-blooded creatures that suckle their young; we are animals with backbones and therefore share a certain basic skeletal structure with such diverse animals as fish and birds. And like all animals, we are dependent upon oxygen liberated by the plant life of the globe.

Where should we begin the search for the origins of man? With the very first life on earth, in the primitive sea where living cells first reproduced themselves. There was no hint then, of course, that more than three billion years later similar cells, multiplied a billion-billion-fold, would manifest themselves in the complex cellular structure of our bodies. Yet it happened. We are here to prove it.

Dazzled, as Shakespeare was, by our own finished beauty, we are apt to overlook the importance of our origins. Yet if we are to gain any real understanding about ourselves, we must learn to recognize the age-old elements from which we have emerged, and how and why they go together as they do. That story, as the next chapter shows, began when life began.

Chapter Two: A Devious Line to Man

It is 3.5 billion years before man will appear on earth. A restless deep blue sea, marked only by occasional ripples of iridescence, rolls over two thirds of the planet. The remainder consists of a single gigantic continent, all brownish rock glinting here and there with patches of bright minerals. For the most part the rock lies flat and naked, but in some areas chains of low mountain ranges rise up and extend for a thousand miles or more; elsewhere, ragged trenches gape across the stony plain.

The forbidding surface is in a continual torment. Almost everywhere volcanic cones and fissures spout dust and vapor or gush crimson rivers of lava that soon harden into blackness. The climate, uniformly tropical and humid, is marked by local fogs, clouds, rain and lightning storms. Wind and waves scour and plow the land. Pale lakes, left by the rains, turn brown with eroded fragments of rock, on their way to becoming soil. A visitor from space would hear a never-ending babble of sounds: the rustle of moving air and the roar of storms, the wash of waves and tides, muffled grindings and explosions as the earth's crust cools and warms with the cycle of night and day.

What he would not hear or see is any sign of life upon the planet. The opaque sea is empty of life; the land shows no trace of green. There is no breathable free oxygen in the atmosphere, which consists mainly of water vapor, hydrogen and two poisonous gases, ammonia and methane. These same active chemicals dissolve and bubble in the waters of puddles and seas. A pitiless stream of ultraviolet radiation, inimical to life, pours in upon the planet from the sun. In such an environment none of the higher forms of life that would later populate the earth could survive an instant. Yet to a space visitor on this hostile planet, this very hostility would be a promise of life to come. For the poisons and turmoil of the primordial earth, oddly enough, are a prerequisite for the appearance of life. It will come in three stages, each stage transforming the earth to bring forth the world that men will live in.

For a billion years, since the birth of the planet itself, the physical constituents of life have been accumulating in the atmosphere and the waters. Now, in the warm primal sea, true life is about to emerge. It will remain in the sea for more than two billion years, constantly changing in its forms and sizes and functions. From the beginning, the changes in form and function will progress inexorably from the simple and primitive to the mysterious complexity of man, following an axiom laid down by the geneticist Theodosius Dobzhansky: "Life tends to spread out and utilize every opportunity for living, no matter how narrow and constraining it may seem to us."

The opportunity for life that existed on the tormented earth 3.5 billion years ago could hardly have been more narrow—but it did exist.

Within the compounds forming the poisonous mixture of the earth's original environment were the elements carbon, hydrogen, oxygen and nitrogen—the basic ingredients of the organic substances making up all living things. In a modern laboratory the four chemicals that filled the earth's early atmo-

Like an army of upside-down ice cream cones, stromatolites, produced by the activity of blue-green algae trapping and binding sediment in layers, stand on a Precambrian sea floor around one billion years ago, in this artist's reconstruction. Stromatolites like these grew as tall as 50 feet, but their odd shapes were determined by a still-inexplicable process.

RELATIVE DURATIONS OF GEOLOGIC ERAS

Geologic Time Chart

ERA	PERIOD	EPOCH	DATE	BIOLOGICAL FEATURES
CENOZOIC	Quaternary	Pleistocene	10,000 to 2 million	First true man: Homo erectus
	Tertiary	Pliocene	2 to 10 million	First manlike apes
		Miocene	10 to 25 million	
		Oligocene	25 to 40 million	First monkeys and apes
		Eocene	40 to 60 million	
		Paleocene	60 to 70 million	First primates: prosimians
MESOZOIC	Cretaceous		70 to 135 million	First flowering plants Last dinosaurs
	Jurassic		135 to 180 million	First birds
	Triassic		180 to 225 million	First mammals First dinosaurs
PALEOZOIC	Permian		225 to 270 million	
	Carboniferous		270 to 350 million	First coniferous trees First reptiles First insects
	Devonian		350 to 400 million	First forests First amphibians First bony fish
	Silurian		400 to 440 million	First land plants First fish with jaws
	Ordovician		440 to 500 million	First vertebrates: Armored fish without jaws
	Cambrian		500 to 600 million	Invertebrate fossils: First shell-bearing animals
PRECAMBRIAN			600 to 4,500+ million	First living things: Algae, bacteria

sphere and seas—water, hydrogen, ammonia and methane—can be made to rearrange their constituent elements to produce the organic materials of life. The experiment is a surprisingly simple one. The mixture is merely heated and exposed to some form of energy —electricity or radiation. It is an experiment that must have taken place countless times during the earth's first billion years. The materials for it were there, in the atmosphere and the water, and so was the energy. There was electricity, in the lightning that ripped through the sky; there was radiation, in the ultraviolet waves that poured from the sun; and there was heat, in the fiery volcanoes that erupted in every part of the earth's crust.

Gradually, the primitive earth's energy and raw materials must have generated the stuff of which life is made—notably the organic compounds called amino acids, which are the building blocks of proteins and also of DNA, the carrier of hereditary patterns for all living things. The sea, particularly, became rich in these materials; modern researchers have called the primitive sea a kind of organic soup. And it was in the sea that the next step took place, at a great turning point some 3.5 billion years ago. Until this point, the raw materials for life had accumulated, but there was yet no life. Then, the great forces of natural energy made some of the available materials join together into new and still more complex substances. Some of these substances had an astonishing capacity. They could reproduce themselves. From the raw materials around them, they could assemble substances just like themselves and proliferate. They were the first living organisms on earth.

We know little about these first organisms. They must have been microscopic and may have resembled modern viruses, bacteria and fungi. They could not have lived by breathing oxygen, for there was no free oxygen to breathe; instead, they got the energy to sustain themselves by breaking down the materials of the organic soup through the chemical action called fermentation, a process still employed by many bacteria and fungi. But since these first living things fed upon the organic materials in which they had been born, they would eventually have exhausted the organic soup of the sea. This was a fatal flaw that sent the earth's original life forms down an evolutionary dead end; in the course of time, as the ecologist Barry Commoner puts it, "Life would have destroyed the condition for its own survival."

Then, about three billion years ago, there came a second turning point, a second opportunity for life. A major waste product of fermentation is carbon dioxide—the bubbles of gas that enliven a fermented drink like beer or champagne. This waste product became the starting point for new forms of life containing the substance chlorophyll. Chlorophyll made possible the process called photosynthesis; that is, it converted carbon dioxide, water and sunlight into sugar, which then became food for chlorophyll-containing forms of life. These forms, freed from dependence upon the ready-made molecules of the organic soup, flourished mightily, slowly evolving into all the varied members of the plant kingdom. What was more important, they in turn produced a third opportunity for life on the earth.

Photosynthesis, like fermentation, has a characteristic waste product. It is oxygen, which over a period of a billion years seeped through the waters in which the first plants grew. The oxygen was lethal to many of the early fermenting organisms, but took another billion years to accumulate in the atmosphere, and then it opened the way for a different, more efficient kind of life. Somewhat less than a billion years ago, certain microscopic forms began to sustain themselves by combining oxygen with living material —from plants or from other forms like themselves. These oxygen-breathing animals, the earliest ancestors of man, soon swarmed in the sea, feeding upon plants and upon one another. From minute one-celled blobs they developed in a fairly short time into highly specialized creatures. Some were mobile and could propel themselves through the water with tiny, whiplike tails; others floated passively or anchored themselves to undersea slopes. Eventually they became sponges, jellyfish, worms and coral.

In a few rare cases, these ancient forms of life have left massive records of their existence. The microscopic blue-green algae, the first plants to evolve, trapped bits of sediment and, layer by layer, built up huge structures called stromatolites, which still exist in smaller versions along the Florida coast. Some very ancient stromatolites resembled nothing so much as upside-down ice cream cones—except that these cones towered as high as 50 feet and were 30 feet wide at the base. One billion years ago they loomed above the silent floor of the sea in greenish white sand-and-algae "forests" that stretched for many hundreds of miles without a break.

The remains of stromatolites, however, are among the few relics of this ancient time. Most plants and animals living then had soft bodies, without the bones, shells and stems that form fossils. Only with the dawn of another era would the fossil record become abundant and comparatively easy to read.

There are three such broad eras in the history of

A cross section of the sea 550 million years ago, during the Middle Cambrian, teems with life. At left foreground, a lobsterlike trilobite called albertella crawls past algae and sponges toward clamlike tritoechia. To their right, an annelid worm inches into a stand of eocrinoids, ancestors of the sea lily. Jellyfish float behind them. In the far right corner, a furry annelid digs into the bottom. Behind it are lingula shellfish, mollusklike hyolithids swimming over sponges, and on the shore, stromatolites similar to those on page 24.

life on earth: the Paleozoic (Greek for ancient life), Mesozoic (middle life) and Cenozoic (recent life). Each era is divided into periods, and some into epochs *(chart, page 26)*. With the earliest part of the Paleozoic Era, the Cambrian Period, a sort of baseline is drawn in the history of the earth—a baseline at which the fossil record of living things begins.

The Cambrian Period has left us an especially rich trove of fossils, in sharp contrast to the Precambrian times, which, except in rare cases, left nothing but stromatolites. The setting had changed little from the Precambrian world. The earth's climate was still tropical and without seasons; the salty sea still cradled a single continent of barren rocks. But the cast of characters was enormous and its members teemed in every part of the Cambrian sea.

Of them all, the most numerous that we know about were the lobsterlike trilobites, which make up

no less than 60 per cent of all Cambrian fossils. Hard-shelled, multilegged, these segmented animals assumed a bewildering diversity of forms. Some had a dozen eyes, some none; some had huge heads, while others seemed to have none. Almost all were small; the largest of them, the giant of its time, measured about 18 inches from blunt head to stubby tail. All are now extinct. But there were other and more promising types of marine life as well. Some left no fossils because they were soft-bodied, like worms and sponges; however, they gave rise to modern descendants. Others were marked by hard shells like those of clams and snails; and still others had casings and forms much like those of modern shrimp.

The calm of the Cambrian ended as a new period, the Ordovician, brought with it floods—the greatest the world has ever known. In seas that inundated the lands, new opportunities for life constantly arose. Seaweeds of every description coated the sea floor with a velvety green cover; the tides streamed through thick forests of slimy fronds; enormous islands of algae grew upon the surface. Organisms that swam or floated near the surface, in the wash of waves and the cycle of sunlight and darkness, developed life styles different from those in the deeper waters of subdued light and even temperatures. In the cold depths of the sea there were creatures that adapted to terrific pressures and also glowed in the dark; on the sea floor, others crept about in the soft detritus that drifted down from above.

Some Ordovician organisms, such as corals, clams, starfish and sea urchins, are still familiar to us in modern times. Others would seem vaguely familiar but disturbingly different in size or appearance. There was, for example, the first true giant of the sea, the nautiloid, a mollusk related to a squid—but the nautiloid was protected by a gigantic hard shell that sometimes grew to a length of 15 feet. Still other animals were insignificant within the Ordovician sea, but they represented potent omens of the shape of things to come. Foremost among them were a few strange, small fish; they were the first animals to possess backbones, the vertebrate structure that today supports all the higher animals, including man.

In the Silurian Period, which followed the Ordovician, fish began to appear in greater numbers, but they were not particularly impressive. These early

The terror of the sea in the Silurian Period, 410 million years ago, a six-foot water scorpion called a eurypterid (below) paddles with tail and hind limbs as it stretches its front claws toward its prey, a school of armor-plated but jawless fish called birkenia (right, above). The fleeing fish swing upward, their tails' lower lobes helping them dart to the surface.

fish, called ostracoderms, seldom grew to more than a foot in length. They had no jaws, only toothless mouths with which they sucked up nutrients from the bottom mud, and instead of true paired fins they grew simple flaps of tissue.

The ostracoderms' skin thickened into a protective suit of heavy, bony armor, and they needed it badly. Relatively harmless creatures such as trilobites and nautiloids were on the decline; powerful and vicious predators were on the rise. Most dangerous of all were the eurypterids, or water scorpions, believed to be ancestors of modern land scorpions. An eight-inch modern scorpion is fearsome enough. But its ancestor, the eurypterid, was the largest animal of the Silurian world—as long as six feet, with a strong paddlelike tail, lengthy claws somewhat like those of a lobster (a six-foot lobster, that is!) and sawlike mouth parts that could rip open the heaviest armor of the little jawless fish. At first glance, it seems almost impossible that the feeble fish could ever have survived when pitted against such a monster.

They not only survived, they prevailed. For one thing, they were vertebrates, able to swim faster and more efficiently than the invertebrate water scorpions that preyed upon them. For another, they were themselves in the course of an evolution that made them still more efficient and effective. Jawless fishes, cruising the bottom and sucking up the muck, had little future; among their few survivors is an unarmored relative called the lamprey, which feeds by fastening its sucker mouth to a living fish. But late in the Silurian Period a new kind of fish, armed with a biting jaw, proliferated. It was only a few inches long, but it was almost certainly a predator, and the harbinger of the next 50 million years of fish evolution.

With the arrival of this superior fish toward the end of the Silurian, there appeared another harbinger of the future—one equally minute but equally important. For the first time, plants began to spread out in herby growths on the shores of the seas. At last, life was about to leave its nursery in the waters and come to land.

Scientists know little about these early land plants; the very evidence for their existence is meager, and they seem at best to have been no more than modest shrublike affairs. But the scientists do know that, in a curious way, the two harbingers are related. During the Devonian Period, which followed the Silurian, the destinies of the first jawed fish and the first land plants proved to be intertwined. Descendants of the plants would become the world's first forests; certain descendants of the fish would become animals that lived in these forests.

This new turning point in the history of life on earth was accompanied by great upheavals in the earth itself. As the Silurian Period drew to its close and the Devonian opened, the crust of the earth crumpled repeatedly and buckled. Slowly, the land was lifted upward; here and there new mountain ranges were formed. Correspondingly, the waters of the earth —and particularly the waters of inland seas— advanced and receded several times. Behind the receding waters were left thick layers of black mud, rich in organic materials that had been accumulating over a period of millions of years and that now were becoming exposed to air and sun.

Never before had the world offered such an opportunity for life, an opportunity as promising and potentially fruitful as that of the primeval sea. In the

Devonian Period that opportunity was seized and exploited. On the shores of estuaries and lakes, in tidal marshlands, wherever the rich mud had been left, plants began to spring up. At first, their foothold was slender. Seaweeds took hold on soil periodically washed by water, then somehow survived the final ebb of the waters that had nourished them. Reproduction away from the protection of water was also chancy. These early plants formed no seeds but had to rely on spores, a rather clumsy adaptation to life on land. A series of complex steps—all taking place only in water or on very wet land—is required to produce new plants from spores. In the presence of moisture, the spores break open to develop intermediate plant forms that are called gametophytes; these gametophytes grow and produce male and female cells; the male cells are mobile, and if one swims along a film of water to a female cell, fertilization will take place—but only if fertilization occurs where nutrients are available can a new generation of spore-bearing plants be born.

Compared to life within the sea, where nutrient-laden waters washed every inch of a plant continuously, life on dry land was hard for plants—at first. But land life had its advantages too. If it offered less water, it provided more sunlight, the driving force of the photosynthetic process that sustains all green plants. Early Devonian plants were small leafless things—inefficient in their use of sunlight—the ancestors of horsetails, leafless shrubs and ferns. But later ones grew true leaves, providing a wider surface to absorb more sunlight—simple slender leaves to begin with, then broad true leaves as the simple ones fused with branches. Because a plant taller and larger than its near neighbors could catch more sunlight, the plants began to grow big and high; and because bigger plants have the special problem of transporting water and nutrients to all parts of their stems and leaves, they developed systems of tubes to circulate moisture and sap. At the same time, new types of root systems probed the soil, hunting for water and nutrients—and, incidentally, anchoring the plants more firmly and permanently.

Eventually, the problem of reproduction was solved too. Before the Devonian ended, some plants probably began to retain their spores in special sacs. In later plants the sacs became part of the seeds, which sheltered the female egg cells and contained stores of foods to nurture embryo plants after fertilization. Though the Devonian Period witnessed only the beginnings of this development, it represents the moment at which plants began to master one of the basic techniques of living on the land. No longer would they need to rely upon lucky combinations of light and abundant moisture to turn spores into gametophytes and gametophytes into new plants. A seed is like a small plant, a partial replica of its parents; once it is fertilized by male pollen—a process that does not require water—it has a good chance of developing to maturity. The seed-bearing plants were successful from the start; their descendants, such conifers as spruces, firs and pines, now make up fully a third of all the world's forests.

To be sure, a Devonian forest looked little like a forest of today. The plants grew in unbroken waves of green upon green; no other color appeared. There were yet no flowers, no pigments that changed colors with the seasons (and no seasons to trigger such a change). These plants were still relatively primitive organisms, simple in structure and green throughout their lives. But if their structures were simple, their outward forms were diverse and often bizarre. Along the ground snaked the plant known as colpodexylon, never more than a foot or two high. Above reared club mosses the size of full-grown trees—mosses like archaeosigillaria, covered with green needlelike leaves. The giants of this forest looked like ferns and are often called tree ferns, but they were not the low-growing ferns common in a modern American forest. They were towering plants, such as aneurophyton, which reached a height of 25 feet, and archaeopteris, which soared to 50.

In one respect a Devonian forest did resemble a modern one. It offered a rich habitat for animal life. In fact, the Devonian forest offered opportunities for new life that far surpassed anything existing on earth today; if ever the term virgin forest was appropriate, it was then and there, not here and now. For hundreds of millions of years the land had been barren while the sea teemed with competitive forms of life. Suddenly, in the space of a few million years, the land became a fertile haven for any animal that would venture upon it. It was there for the taking—untenanted, abounding in vegetable food.

The first arrivals could hardly have been more modest. They were probably such animals as spiders and scorpions, remote descendants of the great scorpions that had ruled the Silurian seas. They breathed air through tubelike structures called tracheae and never developed true lungs—and because their method of breathing becomes ineffective as size increases, they could never grow very large. Not until the end of the Devonian did a single vertebrate animal make its way out onto the land. When the first one did so, it turned out to be a slightly remodeled fish.

Despite the fantastic burgeoning of land plants during the Devonian, this period is generally called the Age of Fishes, and with good reason. In sheer variety and—equally important—in the development of forms with great survival value, the fish of the Devonian surpassed all other life of the time. Not all of them survived, but some of the humblest showed extraordinary staying power. Some jawless fish, for example, swam in the sea at the end of the period as they had at the beginning. And the jawed fish prospered mightily. Gradually shedding their armor and increasing the power of their biting, rending jaws, they developed into the 30-foot dinichthys, the giant of its day and part of a group that probably became the ancestors of the fish we know today.

Among these new fish were the large sharks, skates and sting rays, which form a class of modern fish distinguished by skeletons of cartilage rather than bone. But the rulers of the sea were to be the bony class of fish, and most important among them were the so-called ray-finned fish, characterized by stiff, slightly maneuverable fins. No vertebrates exceed these ray-finned fish in range and diversity. Today they exist in more different types than all other vertebrates put together, their forms ranging from abyssal monsters twinkling with lights to bottom-feeding catfish, intrepid salmon and leaping sailfish.

But in spite of their success, it was not they that carried the processes of evolution along toward man but two other types of Devonian fish then numerous but now mostly extinct. The first to attempt life on land were the lungfish, which developed primitive lungs to ensure a supply of vital oxygen in the changing environment of the Devonian Period. The lungs worked well in two special circumstances—when the supply of oxygen in water diminished, as it did when seas receded and left shallow stagnant pools, and when the water dried up entirely and the fish had to survive in mud. (Lungfish in Africa and South America still use their lungs to survive at such times.) But lungs alone were not enough for life on land. A second group of fish, bearing the jaw-breaking name of crossopterygians, came up with the missing element —fins that they adapted to locomotion on land. These lobe-finned fish not only learned to breathe air but also came to use their muscular lower fins to travel from a dried-out pool to one that still held water.

Gradually, the fins improved as devices for moving on land; the fish first flopped, then floundered, then waddled. Gradually, too, the crossopterygians came to live out of water longer and longer. And finally, they led to an organism with a distinctive new life style. This animal hatched from an egg laid in the water and for a while lived as a fish, absorbing oxygen from water with its gills. But when this tadpole stage ended, its body and life style swiftly changed; its tail and gills disappeared, legs began to form on its sides, it climbed out of the water, it spent the rest of its life breathing the open air and it laid its eggs in water to start the next generation. This cycle is the typical amphibian way of life, a form of existence exemplified today by frogs and toads. By the end of the Devonian Period, fully developed amphibians had made their way onto the land.

The succeeding period, the Carboniferous, was a good time for amphibians to be alive. These were years of relatively gentle changes in the global crust, the land was generally flat, the sea shallow. A slight sinking or uplifting of land, a slight advance or retreat of water could inundate or drain far reaches of the continents. It was a damp world made for creatures like the amphibians, which were at home either in the water or on land. But it also suited some other and newer forms of life.

As the seas advanced and receded, vast areas of the earth stayed marshy, and in these areas the land plants that had come ashore in Devonian times rooted and prospered and spread into the greatest forests of all time. Ferns, club mosses and horsetails flung out their spores, conifers scattered seeds, and in the unchanging, warm climate such plants rose a hun-

The denizens of a late Devonian lake illustrate early and late stages of fish evolution 360 million years ago. At extreme left and at right, onshore, two members of the advanced genus eusthenopteron show their ability both to swim and to wriggle across land seeking water during droughts. Just to the right of the larger eusthenopteron, moving slowly along the bottom, are primitive armor-plated, brownish bothriolepis, which had jaws, and farther right, two of the more primitive jawless spade-shaped fish called escuminaspis. Another early jawless fish, endeiolepis, is shown at far right near the shore. The large fish in the center is the predator plourdosteus, whose snapping jaw could open wide enough to seize relatively large fish. It pursues a more advanced type, the speckled lungfish fleurantia, to the right of which are other fast swimmers, two of the striped fish called cheirolepis.

dred feet or more into the steamy air, first sprouting and then dropping leaves the entire year round. When the great pulpy trunks finally toppled into the brackish swamps, their materials soon decomposed into thick, sludgy layers of peat, to be compressed over the millennia into the coal that provides fuel to power the modern world.

Into the Carboniferous forests exploded a great horde of the world's first insects. They seized the broad opportunity for living offered by the great expansion of greenery, proliferating with astonishing diversity. There were creatures like today's dragonflies, except that their bodies were 15 inches long and they had wingspreads of 30 inches. But the most successful of the crawlers and flyers was—as it still is in the 20th Century A.D.—the cockroach. In that ancient time the earth's surface swarmed with no fewer than 800 species of cockroach, including an enormous one with a body four inches long, and their descendants still live, as many a householder knows, in almost every part of the world.

To the hardy cockroach, modern man owes nothing but grudging respect for its durability. The amphibians, however, must be honored as family ancestors, part of the main line of evolution. These first four-footed creatures reached their high point of specialization and numbers in the marshy woods of the Carboniferous Period, feeding on insects, on one another and—because they were still more agile in water than on land—on fish. They had made themselves at home in inland rivers and lakes, and local droughts were no problem, for they could now waddle easily from one stream or pond to another. Practically all had to moisten their skins regularly,

and all still laid their eggs in water. The amphibian larvae released from these submerged jelly-coated eggs still lived as fish while they awaited their metamorphoses into adults with functional lungs and limbs, ready to test life on land.

At some time during one of the long periods when the sea was receding, one species of amphibian laid eggs that survived a certain amount of drying out. From this species later came a strain whose eggs could resist exposure a little longer. In new generations selective breeding favored those that laid eggs with tougher, more protective shells and still less need for a watery environment. Eventually, a system of membranes developed inside the shell, guarding and serving the embryo, now wombed in its private, minuscule realm of water and food and completing its development within the egg.

The creature that broke out of such an egg was no longer a fishlike tadpole that had to stay underwater until it metamorphosed into an air-breathing, walking creature. Instead it was born ready for land life, a miniature replica of its parents. The newborn infant could run at once after the insects that it already instinctively relished. Such animals were the reptiles, with more efficient backbones, straighter and more mobile legs and better brains than any amphibian.

By the end of the next period, the Permian, which closed the Paleozoic Era of "ancient life" 225 million years ago, the reptiles had overcome the amphibians. The reptiles included large and small plant eaters, and large and small flesh eaters to prey upon the plant eaters. They not only diversified but also spread out to dominate almost every niche of the land. In Antarctica paleontologists exploring ice-free mountain peaks between 1969 and 1971 found fossils of two of these reptiles, lystrosaurus and thrinaxodon, about 400 miles from the present South Pole. The discovery has a double significance. Because the two reptiles apparently also lived in what is now South Africa, their presence in two regions so widely separated reinforces the belief that at the end of the Paleozoic, South Africa and Antarctica were still united as part of the earth's supercontinent. And the abundance and diversity of their fossil remains is testimony to their success as organisms.

Because of their very success, the reptiles of this time, when the era of "middle life," the Mesozoic, was opening, pose one of the major mysteries of modern paleontology. These reptiles were intensively studied during the 1960s, when rich fossil deposits in South Africa were analyzed. But they are only now being fitted into the complex history of life—and the job of accounting for them has produced some surprising results. For one thing, these reptiles were

The amphibians and insects of the Carboniferous Period, 325 million years ago, must have had many confrontations like the one shown here. At left, a 10-inch amphibian, dendrerpeton, cocks a beady eye at a potential meal—the flying insect stenodictya (above), sailing through the air on five-inch wings. The cockroach at right has a less impressive wingspan and flying performance, but nonetheless boasts an outstanding record of survival: as its appearance suggests, it is an ancestor of today's household pest.

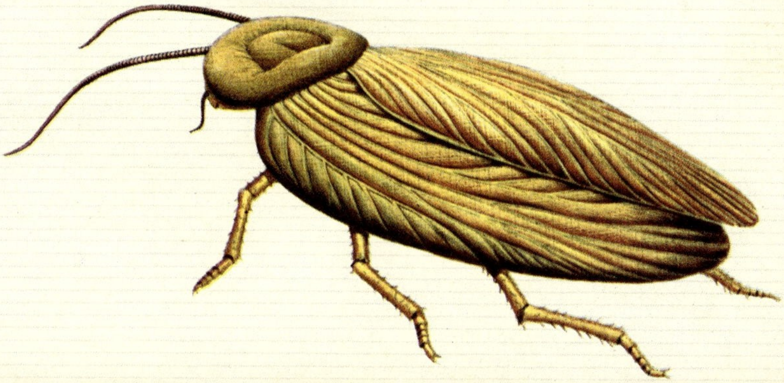

more mammallike—and therefore, in certain respects, more advanced—than still later Mesozoic reptiles that were to drive the mammallike reptiles from the face of the earth. Some of their mammalian characteristics, such as the structures of their jaws, teeth and palates, can be demonstrated in their fossil skeletons. Other mammalian qualities do not show up in fossil remains, but they are presumed to parallel the skeletal similarities; for example, many paleontologists are now of the opinion that these mammallike reptiles, like modern mammals, had either fur or layers of fat under their skins to help keep their blood warm. Paleontologists now agree that before the mammallike reptiles finally died out, they had given rise to true mammals, the form of life that now dominates the earth.

Why, then, did the mammallike reptiles themselves disappear? Here there is less agreement. Part of the answer, paradoxically, may be that a method of controlling body temperature proved to be a handicap to the very reptiles that originated it. During the late Permian Period the climate was fairly cold; a physiological system for keeping the blood and body warm conferred a distinct advantage. On the other hand, during the first period of the Mesozoic Era, the Triassic, the climate warmed up, reducing the value of heat-conserving insulation—and the mammallike reptiles may not have had the other half of the temperature-control system possessed by mammals: a method of cooling the blood and body in warm weather or after heavy exertion. Conceivably, their insulation cooked them inside their skins.

But there were other reptiles around during the Triassic—among them the thecodonts, which had no insulating layers of fat or fur and could radiate excess heat from their bodies with fair efficiency. And the thecodonts had an additional advantage. Their limbs were straighter than those of the mammallike reptiles, a feature that made possible a further advance upon the low-slung waddle of amphibians. Thecodonts walked and ran efficiently on fast-moving, nearly vertical pillars. By comparison, the mammallike reptiles labored over the land with awkward straddling legs; as one authority describes it, they carried their bodies suspended in a "permanent push-up position," in which "inefficient posture added to the problem of temperature control."

The mammallike reptiles, faced by widespread competition in a warm climate, grew smaller and more furtive toward the end of the Triassic Period. Small size made them elusive and also reduced the ill effects of excess heat; because its heat-radiating surface is greater in proportion to its heat-generating bulk, a small animal is able to get rid of body heat more easily than a big one. By the end of the Triassic most of these mammallike reptiles scurried about, no bigger than rats, living on plants, insects and, as rare treats, other small reptiles. Perhaps they were nocturnal in habit; perhaps they lived in burrows or made their homes in hollow trees and rocky fissures, where their enemies could not easily reach them. But these almost-mammals, born before their time, were doomed. Not a single fossil of a mammallike reptile

The mammallike reptile trochosaurus braces to fend off competition before sinking its saber teeth into its prey, the fellow reptile jonkeria at its feet. The most successful land animals during the middle Permian Period, the mammallike reptiles survived over a span of 100 million years (from 300 to 200 million years ago). They were highly varied: jonkeria was a 12-foot plant eater with relatively short teeth; trochosaurus, an eight-foot fang-equipped meat eater.

has been found that can be dated later than the early Jurassic. And yet, although these species eventually died out, they gave rise to other species—ultimately including man himself—which were to outlive the most successful and spectacular reptiles of the Jurassic and the following period, the Cretaceous.

By the time the mammallike reptiles died out, descendants of their archenemies, the thecodonts, had taken over the earth. The thecodonts proved to be a fountainhead of other forms of life. On the one hand they produced such outlandish reptiles as flying monsters, and such persistent ones as crocodiles, which still walk the earth and swim in its waters. On the other, they were the ancestors of all modern birds. The most extraordinary of their descendants were the mightiest reptiles of all time—the ones that made the Jurassic and Cretaceous Periods of the Mesozoic Era a true Age of Reptiles. They were the dinosaurs.

The story of the dinosaurs is so rich and fascinating in itself that it deserves separate treatment; all of the next chapter of this book is devoted to it. The dinosaurs alone, however, do not tell the whole story of Mesozoic life. On the land and in the air and sea, new life forms constantly emerged. The first birds took flight. They were probably about the size of ravens and already had feathers, beaks and a wonderfully efficient system of temperature control. In the sea a mollusk, placenticeras, a relative of today's giant squid with a shell five feet across, could outswim and outfight almost any fish. And on land, plant life took the final step toward its modern forms.

Since Paleozoic times plants had reproduced themselves by scattering spores or dropping seeds. But even a seed plant, the more advanced of the two types, depended upon the wind to carry male pollen from one part of the plant to female egg cells in another part of the plant and thus fertilize the seeds. It was a scheme of propagation more certain than spores but still subject to vagaries of wind. Quite suddenly, during the Cretaceous Period, a new type of

plant appeared—a plant in which pollen and egg cells were brought close together within a single structure, where fertilization was easily accomplished by the gentle swaying of a petal or the brushing of an insect. The structure was a flower, and the flowering plants have changed the face of the earth.

The earliest flowers were probably scentless, and colored a simple green, yellow or white. But they soon began to compete against one another with dazzling colors and rich scents to attract pollinating insects. During the Age of Dinosaurs such modern insects as crickets and grasshoppers flitted among the flowers, and the last of the dinosaurs themselves trod upon dogwood and magnolia, and brushed under the blossoms of laurels, sassafras and palms.

But it was certain reptilian cousins of the dinosaurs that were the strangest of all the evolutionary innovations of the Mesozoic. They invaded the sea to take on its biggest denizens and in the air they far outclassed the fledgling birds.

When dinosaurs dominated the land during the Cretaceous Period, 100 million years ago, their cousin reptiles ruled in the sea and the air. At left, the savage 25-foot-long lizard, tylosaurus, flips its tail—its main swimming organ—as it steers through the water. Above the sea, glide two of the flying reptiles called pteranodon—the largest airborne creatures of all time, with wingspans of 27 feet—ready to swoop down and scoop fish into their gaping, toothless beaks.

The seagoing reptiles flourished in vast shallow inland oceans. One, tylosaurus, grew to a length of 30 feet and had huge hinged jaws equipped with needle-sharp teeth. It was a predator so savage that it could easily kill and eat the bulldog tarpon of its time, a fish that weighed as much as 500 to 600 pounds. Another aquatic reptile was elasmosaurus. Its body was flat and turtle-shaped, though not armored like that of a modern turtle; its legs, originally reptilian, had been transformed into powerful flippers, and its neck was longer than that of any giraffe. A full-grown elasmosaurus might be 50 feet long, and almost half that length would consist of the long sinuous neck.

Competing with elasmosaurus for food was a flying reptile, pteranodon, which soared over the waters of what is now Kansas. Pteranodon was probably the ugliest and certainly the largest creature that has ever taken to the air. It had a wingspan of as much as 27 feet, a long pointed beak and a crest of bone extending backward from its head somewhat like the crest of a blue jay. Though it may have been able to flap its big leathery wings a bit to help itself take off, it was essentially a glider, and fossil remains indicate that it sometimes crash-landed into the sea. Yet it survived for 30 million years, and paleontologists are puzzled by its longevity. Did pteranodon have special adaptations for flight that have not yet been found in the fossil record? Certainly, it had no feathers—but, unlike other reptiles, it may have had a kind of hair or even fur. Some fossils of its cousins show traces of a fibrous material on the skin, and one spectacular Soviet find, which turned up in Kazakhstan in 1966, shows a furry covering, especially dense around the region of the chest. Pteranodon may, then, have moved a certain evolutionary distance toward a temperature-control system comparable to that of a bat or a bird. But for the moment, at least, the 100-million-year-old glider is keeping its secret.

Like the land-living dinosaurs, the swimming and flying reptiles were creatures peculiar to the Mesozoic Era. Like the dinosaurs, too, they disappeared with bewildering abruptness at the end of the Cretaceous Period, the final period of that era. The reasons for their disappearance are still unclear. (For one intriguing hypothesis, see page 88.) But there are several mysteries here, at the dividing line between the Mesozoic and Cenozoic Eras. For one thing, the earth's supercontinent had been gradually breaking apart throughout the Mesozoic and by the end of the era the pieces had become the separated land masses shown on maps today. And not only did the dinosaurs and other reptiles disappear at this time, but with theatrical swiftness as geologic time goes, a host of mammals appeared, and the scene was now set for the slow evolution of man.

No one has described this multiple mystery better or more graphically than noted paleontologist George Gaylord Simpson: "The most puzzling event in the history of life on earth is the change from the Mesozoic, Age of Reptiles, to the Age of Mammals. It is as if the curtain were rung down suddenly on a stage where all the leading roles were taken by reptiles, especially dinosaurs in great numbers and bewildering variety, and rose again immediately to reveal the same setting but an entirely new cast, a cast in which the dinosaurs do not appear at all, other reptiles are supernumeraries, and all the leading parts are played by mammals of sorts barely hinted at in the preceding acts."

Those "mammals of sorts" appeared early in the Mesozoic, only a few million years after the arrival of the dinosaurs. For the next 130 million years they must have diversified toward the varied mammalian tribes that burst upon the world at the dawn of the Cenozoic Era. The earliest of them were probably ancestors of the present-day platypus and spiny anteater—primitive types called monotremes that, like other mammals, had fur and nursed their young with milk, but, like reptiles, laid eggs instead of bearing living young. A second, more advanced group was probably the marsupials, ancestors of the kangaroo and the koala bear, which bore live young. But their young were so small and immature that they had to undergo a "second gestation" in a fur-lined pouch under the mother's belly. However, almost simultaneously with the marsupials, true placental mammals seem to have appeared, laying no eggs and needing no pouch for their young.

The "probably's" in the preceding paragraph tell their own story: The Mesozoic years remain tantalizingly obscure to researchers. Until very recently only a few isolated teeth and jaws of Mesozoic mammals had been found; indeed, the total collection would have fitted nicely into a single shoe box. Scientists could only assume that, as one of them has put it, the mammals of the Cenozoic "came for the most part as migrants from some region not yet studied where they had been evolving even before the last stand of the dinosaurs."

It now looks as though that assumption is correct. During the late 1950s and the 1960s, from discoveries in such diverse regions as China, South Africa, England and North America, a picture has been forming of the thin thread of life that carried the earliest mammals through the long Age of Reptiles. In Lesotho, South Africa, for example, nearly complete fossil skeletons of early mammals were found in 1962 and 1966. (The 1966 find is shown on page 58.) Both finds date from the Mesozoic's earliest period, the Triassic, about 180 million years ago. Both are small —less than seven inches long—and both resemble a modern shrew. The creatures were probably egg layers, like monotremes, but they were certainly mammals and therefore, presumably, more intelligent, efficient and adaptable than the reptiles.

Additional finds like these may bring the answer to the major mystery of the mammals' Dark Ages. Why didn't the first mammals begin at once to take over the world? Somewhere, during the 110 million years that lay ahead, they should have grown larger and more powerful, but they did not. All we know for sure is that these mousy creatures held their own until the dinosaurs died out. Perhaps, like their ancestors the mammallike reptiles, they were slow to develop an effective mechanism for getting rid of their own body heat. (Monotremes, the most archaic of mammals, lack an efficient mechanism to this day.) Perhaps the answer lies in an unexpected quarter —the emergence of flowering plants toward the end of the Cretaceous Period that closed the Mesozoic. It may be that, for their full development, mammals needed the vast range of foodstuffs yielded by these plants—the cereals and grasses, the vegetables and fruits that now feed mammals throughout the earth. Perhaps the dinosaurs were simply too big, too strong and too savage for them to cope with.

Whatever killed off the dinosaurs and their fantastic relatives left a world waiting to be taken over. In the earliest time of the Age of Mammals, the Paleocene Epoch of the Cenozoic Era, the only reptilian survivors were such minor types as crocodiles, lizards, snakes and the ultraconservative, slow-to-change turtles. Out of obscure hiding places crept the mammals, their chance of glory come at last. But what they produced to start with was a curious assemblage of tentative experiments and dead ends. The early creatures were like rough sketches of the highly specialized mammals to come, and for the most part were not even the immediate ancestors of these later mammals. The Paleocene population, small of brain and large of jaw, relatively clumsy and inefficient in feet and teeth, soon passed into extinction, to be replaced by better adapted stocks.

While the experimental models of the Paleocene roamed the earth, however, they did lay down certain patterns for all mammalian life. Among them, for example, were primitive ungulates, or hoofed mammals, which in later stocks produced such familiar grazers as horses, cattle, sheep and goats. From the start the ungulates took to the shrubs, herbs and grasses that thrived in the Cenozoic, but they lacked the special adaptations of the plant eaters that were to come. None, for instance, had the multiple stomachs of a modern sheep, which enable it to graze as fast as it can swallow, then stroll off—or, if necessary, run away—to regurgitate its cud and munch it

at leisure. And they could not have mastered more than a shambling run at best, for none of them had risen to run on their toes like a modern horse or deer. Barylambda, for example, which was typical of one primitive ungulated group, was thick-legged and stout-bodied, as devoid of hair as a hippopotamus, and hoofed only in the sense that each of its 20 toes ended in a heavy hooflike nail.

The meat eaters who fed upon such plant eaters were equally experimental and primitive. Called creodonts (flesh teeth), they came in doglike, catlike and hyenalike forms—but while some had the sharp blade teeth of a modern carnivore, others had oddly blunted ones; in some the claws were pointed and dangerous, in others mere flattened toenails. Not one of them had a brain half as big as that of a modern carnivore of comparable size.

Paradoxically, the Paleocene mammals that were to become the brainiest of all were among the least impressive animals of their time. The progenitors of man and his ape and monkey cousins were by now on the scene. They were the prosimians, which started with a brain the size of a walnut in a body the size of a rat's; even the larger ones, such as plesiadapis, were no larger than a house cat. Fair game for predators, competitors with contemporary rodents, the prosimians scampered around the Paleocene forest, nibbling at palms and sycamores. In appearance and habits they resembled modern tree shrews.

During the following Eocene Epoch their lot—or, to be exact, the lot of their primate descendants—improved somewhat. Foxlike lemurs and huge-eyed tarsiers, more specialized than their prosimian ancestors, survive to the present day, and in Africa and South America the first monkeys of the world made their appearance. At the same time, the ances-

At a Paleocene water hole 65 million years ago, early mammals portend later forms. At top left a cat-sized primate, plesiadapis, scrambles along a branch above the eight-foot-long, root-grubbing barylambda—a primitive forerunner of horses and cows. At lower left a tiny opossum, much like those of today, heads to the cover of a clump of palms. The presence of the big barylambda may protect the two more vulnerable animals from attack by a flesh eater, the wolverinelike oxyaena, poised at right below.

tors of such modern mammals as the camel, the horse and the rhinoceros ventured forth, although in forms that would be almost unrecognizable today. For one thing, they were astonishingly small—the camel the size of a rabbit, the horse hardly larger than a fox, the rhinoceros only as big as a dog. For another, their bodies lacked many of the marks that now distinguish them: the camel was humpless, the rhino hornless, and the horse almost hoofless. All displayed more promise than performance.

During the Oligocene Epoch the promises began to be realized. In the matter of size, for example, some of these animals more than made up for any handicap they had started with. The rhinoceros, particularly, produced Oligocene giants unparalleled in mammalian evolution. One of them, baluchitherium, measured 25 feet long and 18 feet high at the shoulder—the largest land-living mammal of all time. But the greatest realization of earlier promise was achieved by the primates, the order to which man belongs, for during the Oligocene they made the most important forward leap of evolution.

The rapid progress of mammalian life during the Oligocene has only recently been revealed by excavations in the Fayum Depression of Egypt, which has turned out to be a bonanza of fossils. The finds include one fossil that is among the most important links in the human line to be found in many years. Today the Fayum is a desert basin on the eastern rim of the Sahara, but if time could be rolled back 40 million years to the Oligocene, the scene changes dramatically. Here is a humid tropical forest, crowded with fan palms and papyrus and teeming with mammals—familiar yet oddly unfamiliar—that foretell the creatures we know today *(painting)*.

The huge arsinoitherium, nearly 10 feet long and

A mild monster of the Oligocene Epoch in North Africa, 10-foot arsinoitherium backs off on a bank as two moeritheriums suddenly rise out of water; the spectator in the tree in the foreground is aegyptopithecus, a primate. The semi-aquatic moeritherium was related to the elephant; the horned arsinoitherium, though of uncertain lineage, clearly resembled a rhinoceros—and the aegyptopithecus may have been the common ancestor of apes and men.

6 feet high at the shoulder, is clearly rhinoceroslike —but it bears one of the wickedest sets of horns since the days of the dinosaurs. The 10-foot-long semiaquatic moeritherium looks somewhat like a gigantic dachshund—but has long incisor teeth and an elongated upper lip; it is actually a relative of the elephant, and in future years the incisors of its descendants will become tusks, while the upper lip will fuse with the nose and grow into a trunk. Yet it is not these monsters but rather a much smaller frightened creature shinnying up a tree trunk that deserves the most attention. For this insignificant-looking animal, aegyptopithecus, may prove to be the common ancestor of apes and men.

Aegyptopithecus is the great find of the Fayum. Before 1960, its very existence was unsuspected, for only seven bits of primate bone had been found in the Fayum. Since then, largely through the work of Elwyn Simons of Yale University, hundreds of fossils have been turned up, and in 1967 an aegyptopithecus skull was found virtually intact. In life the animal was an ape, roughly the size of a gibbon, with teeth much like those of a gorilla. It was certainly an ancestor of modern higher apes, and if it turns out to be the long-sought common ancestor of both apes and men, it will fill in the most intriguing gap in the story of primate evolution during the Oligocene Epoch.

By the close of that epoch, 25 million years ago, the Cenozoic Era was more than half over. Only three more epochs, all of them short, remained to set man in place and bring the earth's life down to the present day. During the Miocene Epoch, a feeble apelike creature called Ramapithecus branched off from the ape line. In the Pliocene its descendant, Australopithecus, became a man-ape, a borderline being who connects humans to their nonhuman past. And in the Pleistocene, the epoch in which we live, true men arose and flourished. All these developments took but 15 million years—the wink of an eye in a world in which the story of life began 3.5 billion years ago.

Record of the Fossils

Man's world contains a treasury of prehistoric evidence. It is found in rocks, and it consists of fossils. A fossil is any trace of ancient plants or animals—from the imprint of a leaf to the skeleton of a giant vertebrate—that has been preserved in the earth. The fossils are often found in rock that has built up on a seabed: silt has drifted down to the bottom, covering the remains of an animal or a plant. Then centuries later, when the sea has retreated, the rock containing the fossil is exposed and can be dated. This essay begins with a fossil some 10,000 years old *(left)*, and proceeds two billion years back in time.

10,000 to 12,000 years ago. *The skeleton of this Irish elk, megaceros, complete with nine-foot antlers, was preserved in a peat bog.*

30 million years ago. *This birch leaf fell into an Oregon lake and drifted to the bottom. The weight of successive layers of silt settling on it compressed it into a thin film of carbon, which kept intact the outlines and the surface features of the living foliage.*

40 million years ago. The butterfly below, prodryas, was covered by ash during the eruption of a volcano in Colorado. As the ash hardened into shale, it fossilized not only the major external parts of the butterfly's body but even the delicate markings on its fragile, near-transparent wings.

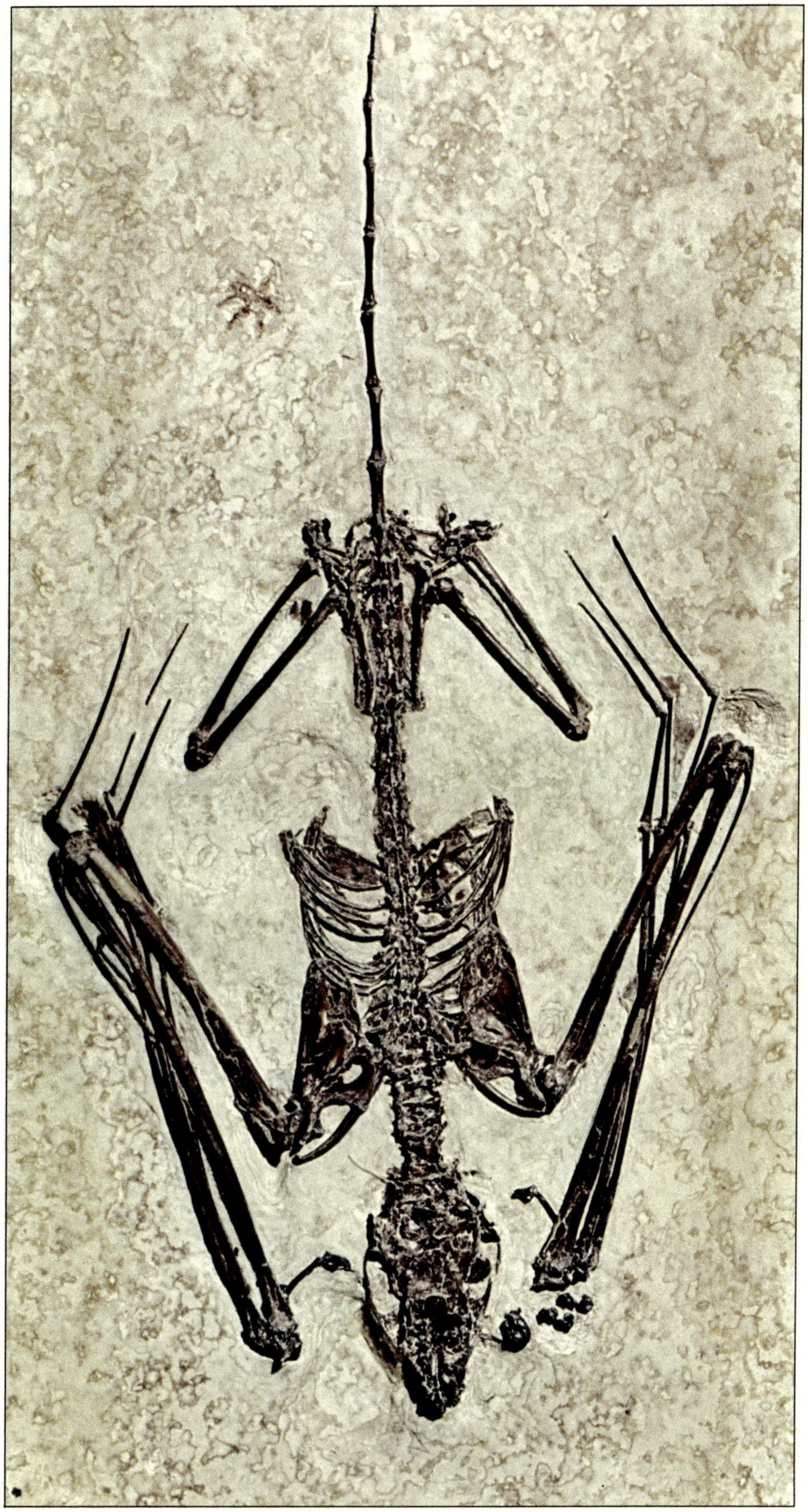

45 to 50 million years ago. *The skeleton of a male bat, this fossil was found embedded in marl—clay mixed with calcium carbonate—in Wyoming's Green River. It is an extraordinary remnant of an extinct creature that was five inches long, had a 12-inch wingspan and evidently was a fish eater. The fossil still has remnants of its delicate wing membranes, cartilage and bones as thin as human hair.*

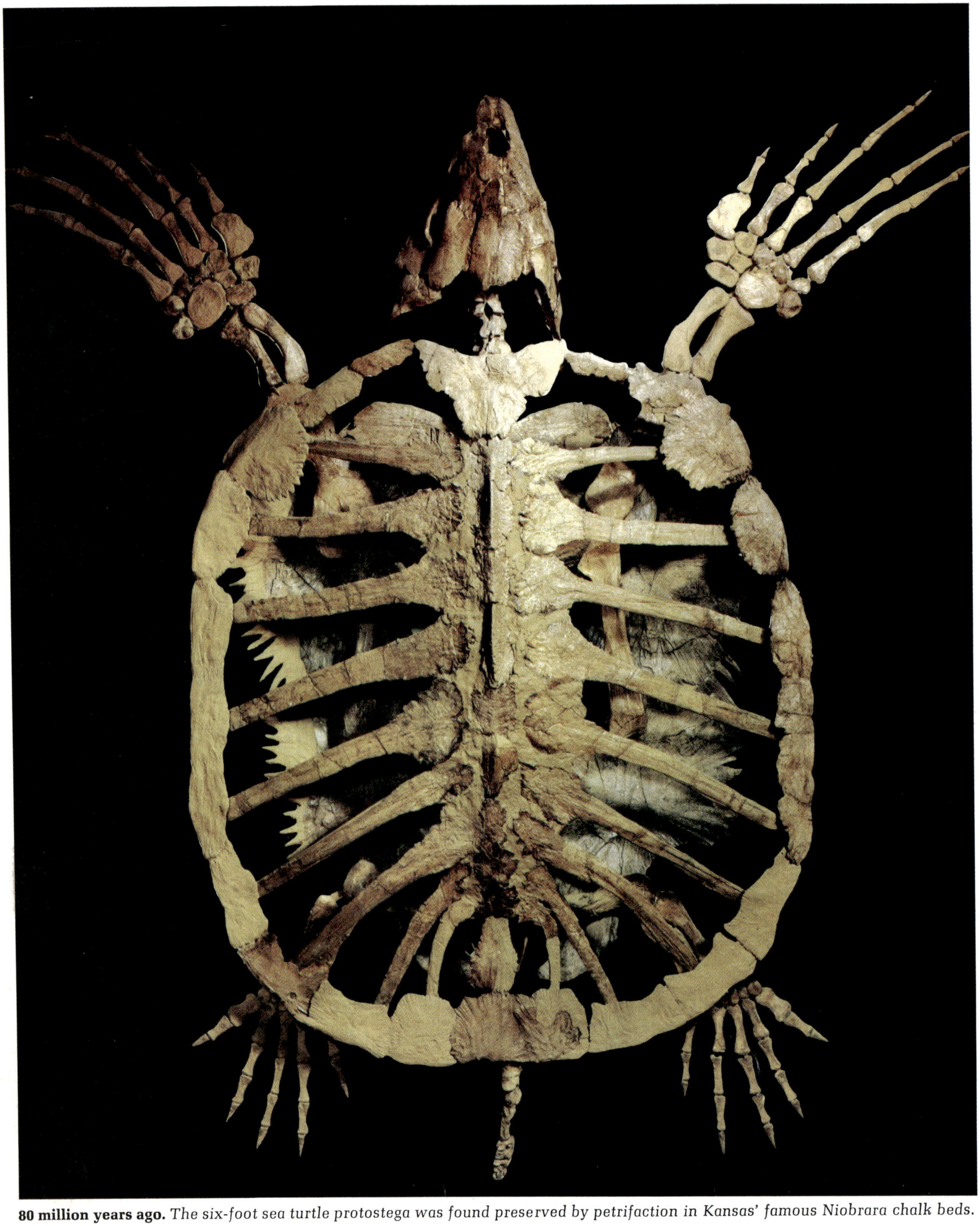

80 million years ago. *The six-foot sea turtle protostega was found preserved by petrifaction in Kansas' famous Niobrara chalk beds.*

100 million years ago. *This community of crinoids (sea lilies) sank into the bottom of a muddy Kansas lagoon. The mud turned to limestone, preserving the hard plate encasing the crinoids' bulbous bodies and reedlike arms.*

100 million years ago. *An ant in amber, this worker from the species sphecomyrma was trapped in the resin that trickled down a tree trunk in New Jersey. As the resin hardened into amber, it kept intact the ant's external skeleton but not its internal organs.*

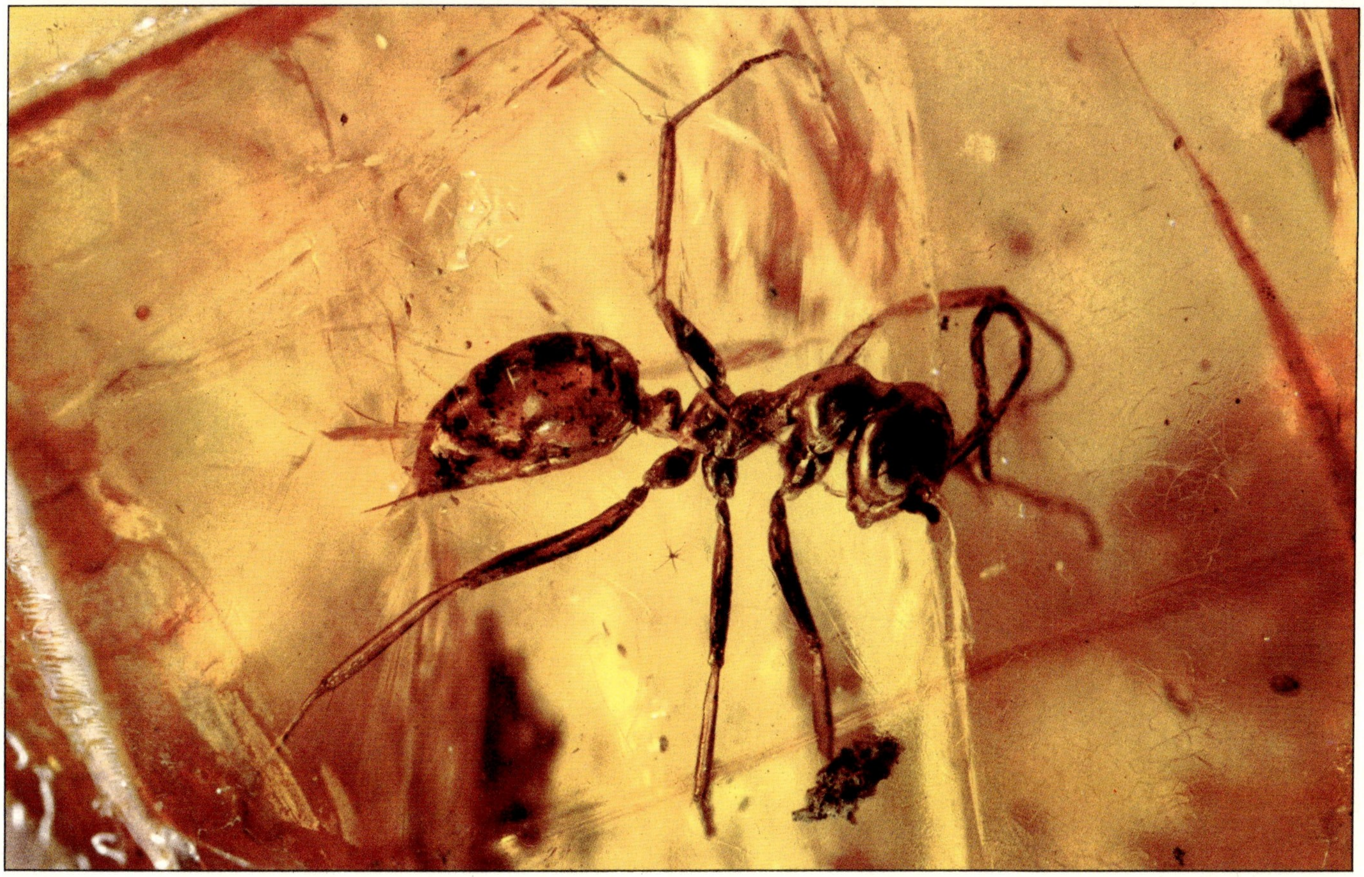

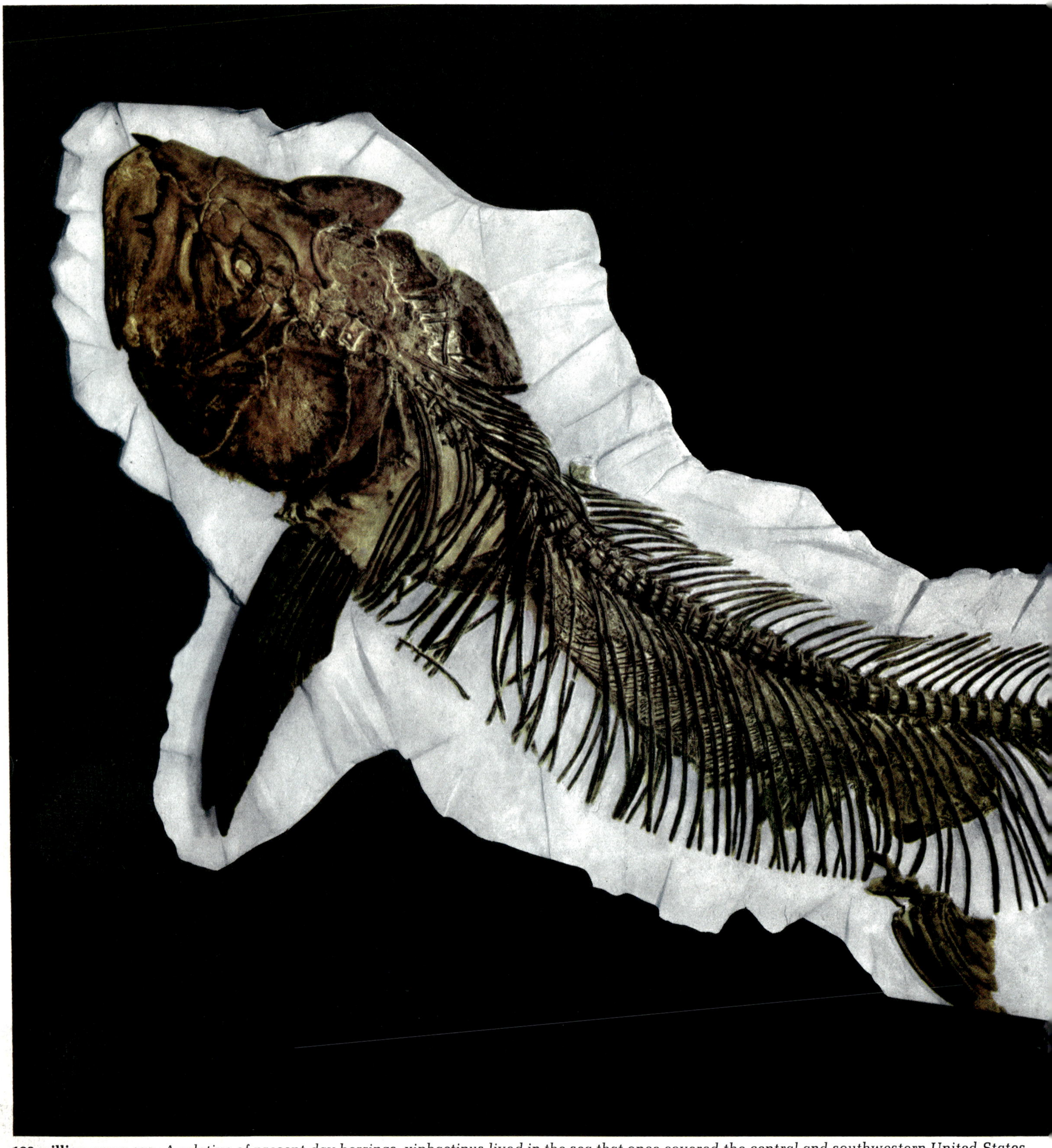

100 million years ago. *A relative of present-day herrings, xiphactinus lived in the sea that once covered the central and southwestern United States.*

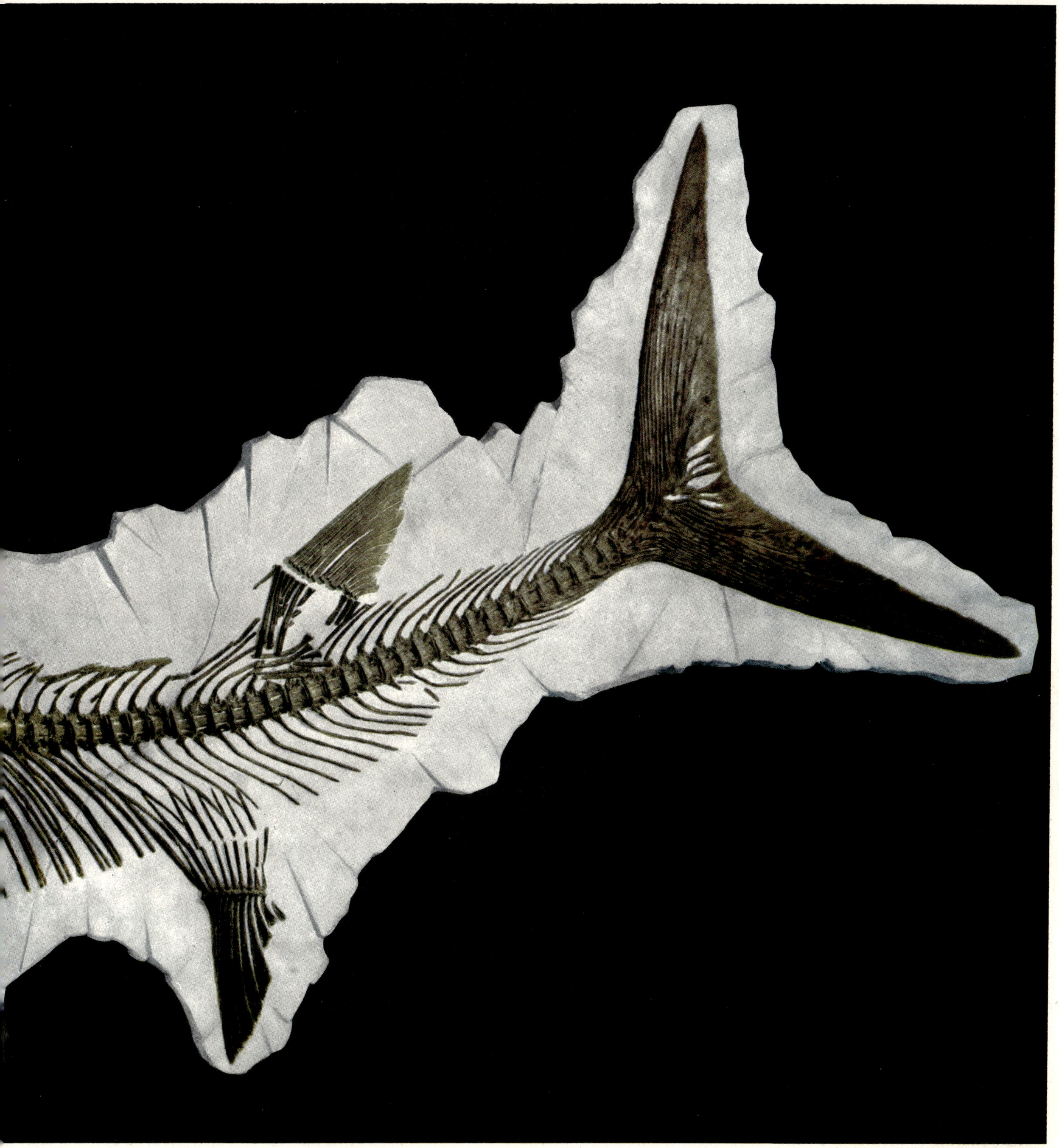

Its 14-foot skeleton was found in a chalk formation near Austin, Texas, with its final victim, a four-foot-long ananogmius, under its front ribs.

150 million years ago. *Pterodactylus, a three-inch-long flying reptile, fell to the bottom of a sea in Bavaria. Because the sea floor was poisonous, the body was not eaten, and its bones remained intact in a limestone matrix.*

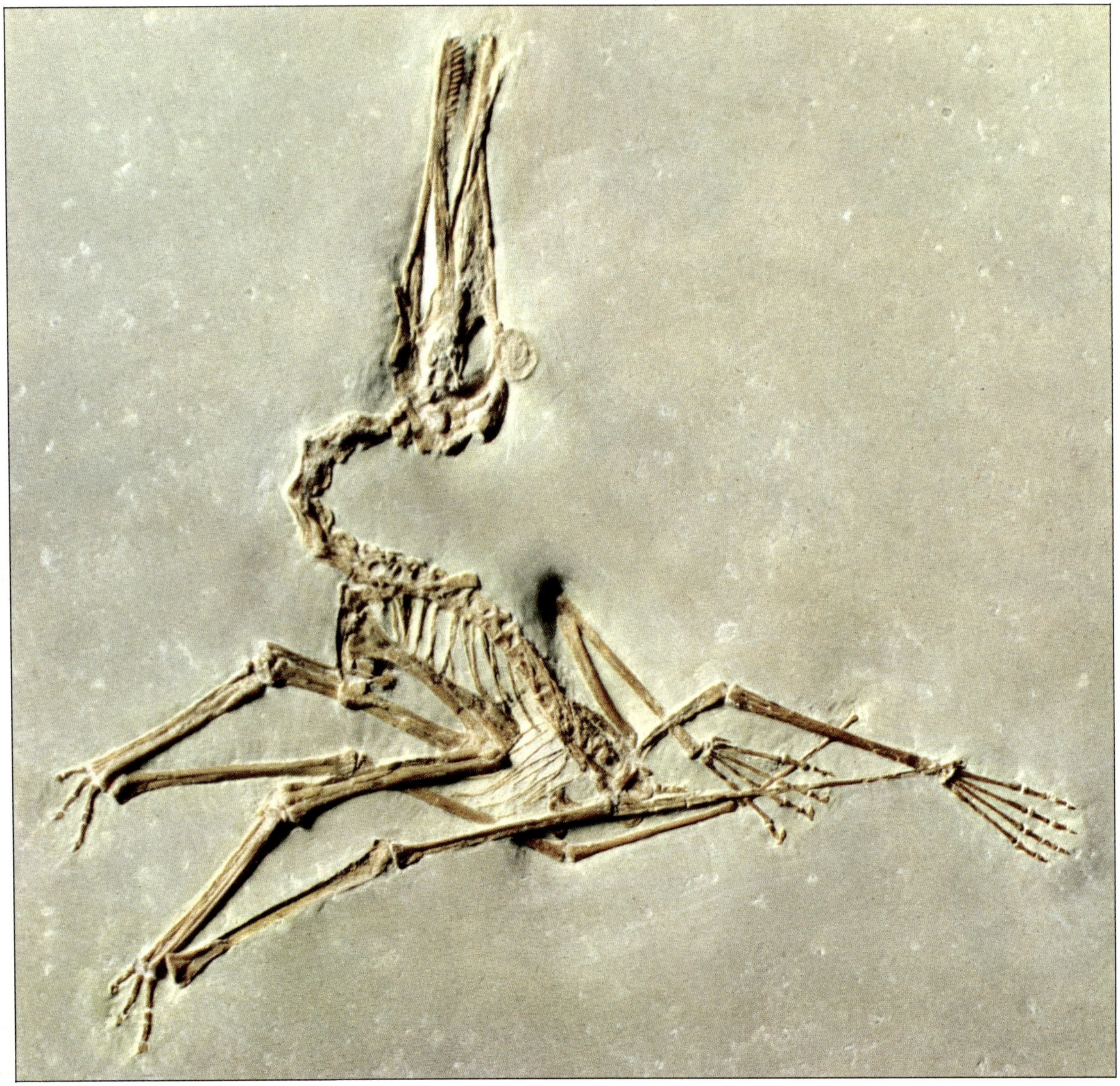

135 million years ago. *This ancestral horseshoe crab, mesolimulus, died lying upside down in a German lagoon. Porous limestone formed over it, dissolving its soft internal organs but leaving natural molds in their place.*

165 million years ago. *This leaf of the plant known as zamites comes from France, where the featherlike outline of its compound leaf had been preserved by compression. Now extinct, the zamites belonged to a group of plants, the cycadophytes, that had wide distribution in the Age of Reptiles.*

180 to 185 million years ago. *The mammal megazostrodon, a smallish quadruped that may have resembled a shrew, lived in southern Africa. This six-inch skeleton, found in red siltstone, has a fairly complete skull.*

190 million years ago. *The primitive bony fish semionotus died in brackish water in New Jersey, where its remains were found embedded in finely grained shale. When dilute acid was poured on the remains, it ate away everything but the shale, in which was left a perfect mold of the fish.*

225 million years ago. *Thrinaxodon, a foot-long mammallike reptile from South Africa, is a famous fossil known to paleontologists as Baby Doll. Its bone cavities have filled with minerals, forming an exact stony replica.*

280 million years ago. *The extinct seed fern neuropteris, which grew to be 20 to 25 feet tall, was not a true fern. But as the design of this beautifully detailed frond found in an Illinois coal bed indicates, its foliage was extremely fernlike. The specimen was preserved through compression.*

370 million years ago. *The rugose, or wrinkled, coral shown here is really a colony of scores of fossilized coelenterates—tiny marine organisms that secreted calcium carbonate to form hard exteriors around their soft bodies. It was found in Indiana.*

500 million years ago. *An early ancestor of the shellfish, the one-and-a-half-inch trilobite below was buried in sediment in a Czechoslovakian lake. As it decomposed, the iron sulfide known as pyrite formed out of the decaying matter and sediment, and gradually fossilized the trilobite.*

Two billion years ago. *These stromatolites from Minnesota are hardened sediment once bound by now-decayed blue-green algae.*

Chapter Three: Nature's Grand Failures

In a Princeton museum, 30-foot antrodemus, a dinosaur that became extinct 135 million years ago, rears its reconstructed head.

Every great parade has its casualties: the champion runner who fumbles the baton in an Olympic relay race; the aged cardinal who has a heart attack during an august assembly in Rome; the Boy Scout who drops the troop standard and retires in misery from an international jamboree.

So it has been through the billions of years of the grand procession of life on earth. Thousands of forms have arisen to swim, wriggle, crawl, walk or fly past some immutable reviewing stand and then collapse. They include creatures too tiny to be seen by the human eye, and they include the most enormous ones that ever walked the earth. Some lived briefly and insignificantly, but others lorded it over their fellow creatures for 135 million years before they too died out. When they disappeared it was as though they had never been. Most of their successors to power never saw those that earlier had come and gone, so separated were they by the millennia.

Certainly no creature in this parade of life, each taking the lead in its turn, was able to deduce that there had been others. It was left to man, a very late starter in the parade, to learn about those that had gone before by probing among ancient stones for fossil shadows of vanished creatures.

Some dropouts were crucial to the development of man himself, whose line can be traced to early vertebrates that first possessed a rudimentary backbone and the beginnings of a brain. Among these ancestral creatures were the ostracoderms, tiny armored fishes two to six inches long whose heads, bodies and tails were covered by little plates of bony armor. Other early forms were mere bizarre experiments, like those monstrous mammals of 45 million years ago, the titanotheres. They evolved into five-ton battering rams up to 8 feet tall and 15 feet long. On each snout was a large appendage of horn resembling an upside-down coat hanger. The function of this growth is not clear; but fossil titanotheres have been found with badly broken ribs, and possibly the males ran at one another in combat during the mating season.

No extinct creature so titillates the imagination, however, as do the great failures of the Mesozoic Era, a period of time from about 225 to 70 million years ago. The Mesozoic is known as the Age of Reptiles, but these creatures were no garter snakes or lizards or even crocodiles. They were reptiles that soared through the air on leathery wings wider than those of the Wrights' first glider. They were reptiles that lurked in the dark waters and had 25-foot-long necks attached to round bodies, like snakes glued to turtles.

And best of all, there were the dinosaurs. There were all kinds of dinosaurs. Some chased one another and fought savagely. Some ate shrubbery, and some ate those that ate shrubbery. They wolfed down one another's eggs and babies, left footprints big as washtubs and scared the wits out of the little emerging mammals, whose potential both mammals and reptiles failed to appreciate. Then, all of a sudden, the dinosaurs were gone.

For 135 million years the dinosaurs ruled the earth by their skill and power. In the past century they have threatened to take it over again, by charm alone. No creature of myth or magic, of nightmare or dream, including such fanciful forms as the mermaid and the dragon, the gryphon and the unicorn, has captured the imagination of modern man as has the dinosaur. No man ever saw a living dinosaur, not even Australopithecus, the ape-man link to man's animal predecessors. No man had dreamed of such a beast

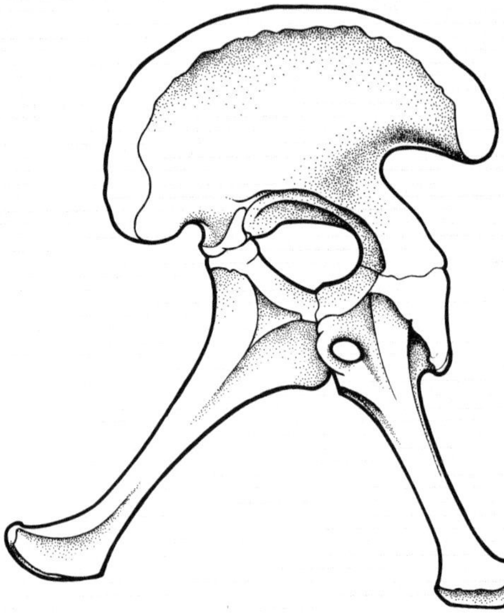

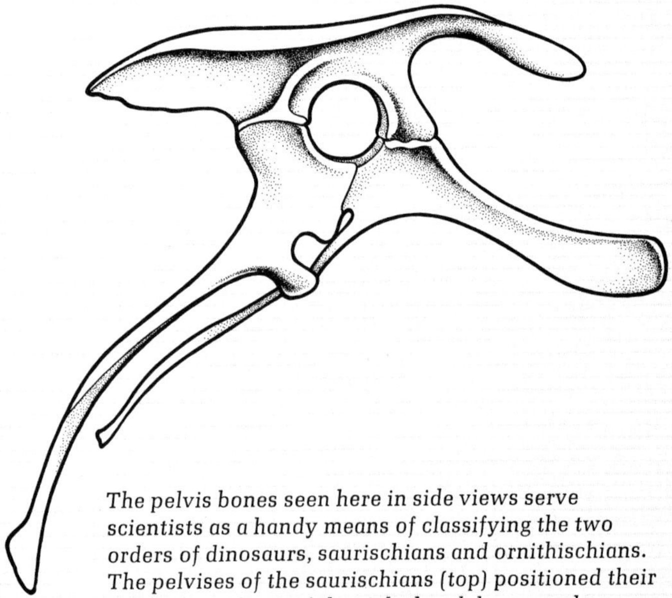

The pelvis bones seen here in side views serve scientists as a handy means of classifying the two orders of dinosaurs, saurischians and ornithischians. The pelvises of the saurischians (top) positioned their stomachs in front of the right-hand downward thrusting bone, called a pubis. The ornithischians had a better arrangement in the pelvis shown at bottom: The pubis had a slender rodlike section extending back toward the tail, while growing forward, toward the right, was an almost horizontal section. (This section actually was divided, "V"-like, in the dinosaurs but only one side of the "V" is visible in the drawing.) Below and between the reshaped pubes was room for an enlarged stomach, an advantage for the vegetarian ornithischians, which had to eat more food than a carnivore to obtain equivalent amounts of nourishment. The saurischian group of dinosaurs included both vegetarians and carnivores; somehow the vegetarian saurischians do not seem to have been handicapped by their less efficient pelvises.

until a little over a hundred years ago. During the 19th Century, a time of great fascination with geology and natural history, so many fossil remains turned up in England that a British scholar named Richard Owen coined the name *Dinosauria* from the Greek *deinos* (terrible) and *sauros* (lizard). His scholarship and his Greek were impeccable, but this scary description never seemed appropriate. Once man had "remembered" dinosaurs, he was bewitched.

When fossil remains began to turn up in 19th Century North America, the dinosaur rush assumed almost the proportions of a gold rush. A park was set aside to preserve dinosaur remains, concrete effigies were raised on Western hillsides, skeletons were dug out and reconstructed. In our day oil companies use dinosaurs in advertising, cartoonists make jokes about them, comic-strip heroes ride them like horses and generations of schoolboys have amazed their mothers by pronouncing such tongue-twisting names as stegosaurus, ankylosaurus, diplodocus.

The dinosaur has been elevated, in the human heart, to the status of honorary mammal. And in a reversal of the normal course of events, science has in the past decade been following the human heart: more and more paleontologists are ascribing to the great reptiles such sophisticated characteristics as agility and mobility, a capacity for great expenditures of energy, and a basic posture more like that of advanced mammals than of reptiles. It is time for another look at the dinosaurs. Primitive mammals shared their world but never dominated it. Only when the terrible lizards died did the spectacular rise of the mammals—and therefore of man himself—begin.

While dinosaurs lived they made up a huge family —an estimated 250 different kinds have been iden-

A Diversity of Dinosaurs

Extending 90 feet along its backbone from nose to tip of tail, a 150-million-year-old skeleton of diplodocus, reproduced in the model at right, shows the great size of one of the largest of all dinosaurs. In its bony bulk, the skeleton is palpably real—but difficult to visualize as it appeared when clothed with flesh. Yet the way these long-vanished reptiles looked when they were still alive can now be reconstructed by scientists. Assembling the bones for a skeleton indicates its general size and shape, and paleontologists can then piece together other scraps of information—measurements of the size of fossil footprints, the positions of the bones when first discovered —to arrive at a fairly precise estimate of the creature's length and weight during its lifetime. And after studying such large animals of today as elephants, they can guess at how diplodocus walked and how his flesh parts attached on

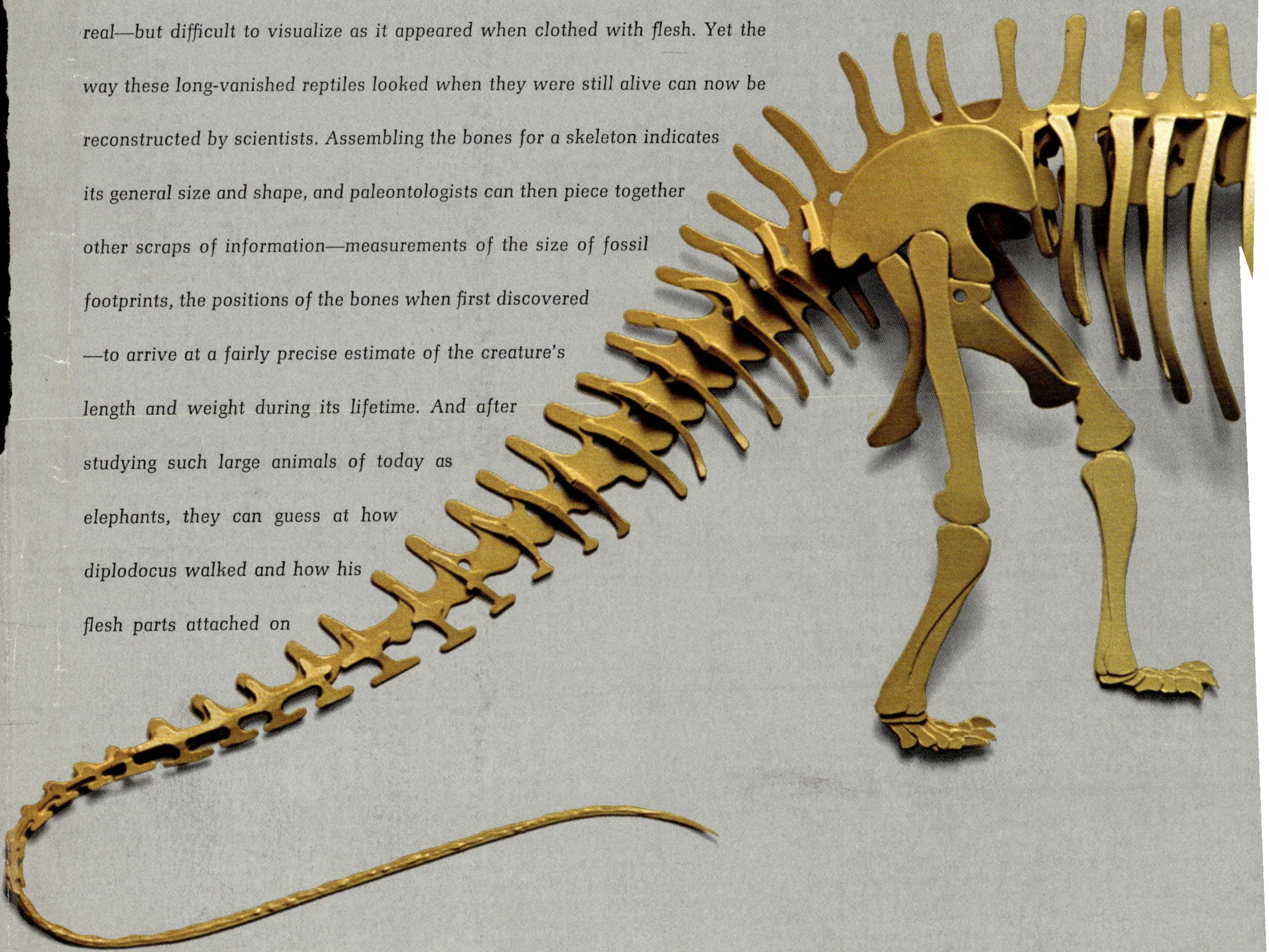

The dinosaurs of the time periods shown in this chart are grouped below under two distinct orders —saurischian and ornithischian —and color-keyed to identify seven suborders. The thecodonts (extreme left) are thought to be the dinosaurs' forebears.

Thecodonts
(*herbivorous and carnivorous*)

Saurischians
Theropods (*carnivorous*)
Prosauropods (*herbivorous*)
Sauropods (*herbivorous*)

Ornithischians
Stegosaurs (*herbivorous*)
Ornithopods (*herbivorous*)
Ankylosaurs (*herbivorous*)
Ceratopsians (*herbivorous*)

PERMIAN PERIOD
270 million to 225 million years ago

TRIASSIC PERIOD
225 million to 180 million years ago

JURASSIC PERIOD
180 million to 135 million years ago

Compsognathus 3 feet 15 pounds
Allosaurus
50 feet
16,000 pounds
Brachiosaurus
80 feet
110,000 pounds
osaurus
eet
00 pounds
Brontosaurus
90 feet
70,000 pounds
Scelidosaurus 12 feet 275 pounds

tified, and new kinds continue to be uncovered. There are two main types: saurischian, with a triangular hip structure resembling that of such modern reptiles as crocodiles; and ornithischian, with a rectangular hip structure like that of birds. Both groups had creatures of all sizes. Some dinosaurs were small as chickens. Some were tall as four-story buildings if they stretched their necks. The heyday of the biggest was toward the end of the middle Mesozoic Era, a period called the Jurassic.

What was it like then? Today, 140 million years later, we know in surprising detail. When the sun rose over the dinosaur homeland now called Colorado and Wyoming, it illuminated, not the 7,000-foot-high plateau of the present, but low, moist plains. On the western horizon, brightening slowly in the morning light, were the silhouettes of hills, the beginnings of the Rocky Mountains rising along the Pacific coast. The coast itself was still under water.

From Montana to New Mexico the area looked like some regions of Panama today. There were dense forests interspersed with dry uplands. Sluggish rivers running down from the west carried loads of silt to form swamps and deltas, lakes and ponds. The weather was mild, with little temperature change between day and night, summer and winter. Vegetation was lush. It was an intensely green and brown world.

It was also strangely silent. There was no bird song. As dawn broke, flies and beetles might have been visible marching up and around the rotting stumps of giant trees. Among the ferns that grew thickly at water's edge and crept across the moist stones, gnats and dragonflies moved. There might be a noise, small and sharp, as a tiny four-legged insectivore, a primitive mammal, scurried through the underbrush in a dash after a centipede.

Busily eating, the little insect eater could neither see nor hear the creature that had it for breakfast. Its nemesis was ornitholestes, a small dinosaur built like a bird but with very sharp teeth in a reptilian skull. This dinosaur, which grew to about six feet, walked on hind legs with its tail stretched out behind, and its front limbs had sharp, tearing claws to pin its prey while the teeth went to work.

If there could have been an observer watching ornitholestes finish breakfast, he would have noticed another stir in the warming morning. A monster was abroad in the neighborhood, and dozens of small creatures quivered with alarm as the earth shook under them and the thick, brittle leaves of palmlike cycadophytes swayed with a dry rasping sound. The monster was brachiosaurus, at 55 tons the biggest land animal of this or any other era. It lumbered through the underbrush mindless as a bulldozer, great elephantine feet crunching down on shrubs and plants, heavy tail carried slightly up from the ground, and thrusting neck reaching 40 feet into the tops of the cypresses and pines.

Brachiosaurus was a monster, but it was not a meat eater. Its small head was equipped with peglike teeth at the front of the jaw, useful only for snatching off bits of branches, bark and leaves. The beast spent most of its days eating, sometimes accompanied by other vegetarian dinosaurs to which it was related. They did not quarrel as fiercely over food as did the meat eaters, but they got quite rowdy with one another over good grazing territory, and they could reduce a jungle to shreds in a matter of days.

It took enormous amounts of food to fuel those

Text continued on page 72

huge bodies, and soft plants and shoots could not provide sufficient nourishment. Brachiosaurus ate tons of leaves and soft bark; occasionally in its haste it must have got a mouthful of wood or rotting fiber by mistake and spat it out like a tobacco-chewing mountaineer. The mechanical breakdown of all this raw material into energy was carried out not in the mouth but elsewhere, possibly in a powerfully muscled stomach and perhaps with the aid of pebbles like those in the gizzards of fowl. Big "stomach stones," polished by digestive acids, have been found in a few herbivorous dinosaur fossils.

Brachiosaurus differed from its fellow herbivorous giants primarily in the size of its front legs. They were longer than the hind legs, and won this dinosaur its name, which means "arm lizard." Its companions in those days in Colorado and Wyoming were the almost equally ponderous *Brontosaurus* (thunder lizard) and the longest of all, the 90-foot diplodocus. Both of them, like brachiosaurus, walked on all fours although their hind legs were longer than their front legs. They had scaly, reptilian skin and probably were a light to medium color that served as camouflage against predators. Whether or not the big dinosaurs made any noise, apart from the crash and crunch of breaking shrubbery and the watery squish of their great feet along the shoreline, is not known. Of modern reptilians only the crocodile is noticeably noisy, and even its repertoire is small. It may be that the giants of 140 million years ago were silent.

Toward noon, the Jurassic sun burned down and brachiosaurus, brontosaurus and diplodocus ambled into a patch of shade. Lesser reptiles also headed for shade, slithering out from under the massive feet of the dinosaurs to find shelter among the small horsetail plants at the water's edge. There was very little movement except in the water. Primitive fish swam there, and now and again from a low perch a small type of pterosaur, an ugly, batlike flying reptile, let go its toehold and glided out to snap at an unwary fish. These pterosaurs were about 24 inches from long, tooth-filled skull to tail, and their wings, like bat wings, were strips of leathery membrane attached from one elongated finger down along the body to the short, gripping feet. Some pterosaurs may have been capable of brief powered flight, but they were essentially gliders, drifting out over the water and then catching a rising air current to return safely to a low shrub or tree. From such a perch a pterosaur could use its clawed feet to climb to another branch high enough to launch another glide.

The pterosaur's chief competitor in the fish-catching trade was *Ichthyosaurus,* called the fish lizard because it looked like a swordfish. It generally grew 10 to 15 feet long, but ichthyosaurus remains more than 40 feet long have been found in Nevada.

Perhaps brachiosaurus, temporarily full, even dozed off briefly in the shade. But in that distant time it might not have been long before another thumping of the earth announced the arrival of allosaurus, a ravening meat eater. From big, ugly snout to heavy tail, allosaurus measured 50 feet and weighed eight tons. Seen in repose it somewhat resembled a king-sized kangaroo, but it was not. Instead of hopping, it walked on hind legs, tiptoe fashion, on its massive toes. And it ran, leaning forward from the hips and using its sturdy backbone to hoist its heavy tail almost horizontal for a counterbalance. Yet the head, not tail or legs, was the business end of allosaurus. It was two feet long with very large eyes and quite ad-

CRETACEOUS PERIOD
135 million to 70 million years ago

Dromiceiomimus
10 feet
175 pounds
Tyrannosaurus
50 feet
20,000 pounds
Spinosaurus
45 feet
14,000 pounds
Parasaurolophus
30 feet
10,000 pounds
Edmontosaurus
40 feet
12,000 pounds
Triceratops
35 feet
20,000 pounds
Leptoceratops
7 feet
1,200 pounds
17 feet
8,000 pounds

The bones of diplodocus, a dinosaur common about 150 million years ago, are re-created in a scale model based on a skeleton found in Utah.

to his skeleton. These principles have been used to construct the illustrations inside these pages. Together, they make up a display of paper sculptures that were built to scale and were then photographed. The chart shows a sampling of representative dinosaurs from their immediate ancestors, the thecodonts, up to the grotesque, armored monsters that evolved last, closing out the giant lizards' long heyday; the parade is drawn up with all dinosaurs of similar ancestry keyed by color.

Lengths and weights given with each picture are based on estimates of adult dimensions. Seeming discrepancies between sizes and weights arise not merely from the astonishing diversity of these creatures but also from the application of simple mathematics. Since the weight of an animal increases roughly as the cube of its length, 60-foot-long euhelopus, though only twice as long as 6,000-pound albertosaurus, weighs approximately nine times as much. In the case of the 7,000-pound sauropelta, its extra weight comes not so much from its protective armor of bony plates, which were comparatively light, as from its internal structure; the paleontologists' findings indicate that its body consisted mostly of solid fat.

equate nose and ears, but its real purpose was to support massive jaws. Almost all of the lower half of the skull was jaw, lined with sharp, thin-bladed, three-inch teeth. The upper skull was lightly built and loosely attached so that it could give a little, permitting allosaurus to swallow enormous chunks of meat.

Allosaurus was almost always on the prowl, stepping along carefully with a gait like that of a chicken strolling in search of food in a farmyard. Once it spotted tasty prey, such as the other dinosaurs resting in their ancient landscape, it made a short rush, opened the great jaw and sank those awful teeth into a chunk of saurian flank. Its bite was so big that allosaurus generally put both of its front feet onto the victim to gain leverage to tear loose the mouthful. But the plant eaters were easily alarmed and well equipped for defense. They lashed long, whiplike tails at an attacker. If they missed, the predator probably got away, for many meat eaters could outrun vegetarians.

One dinosaur with unique equipment for defense was stegosaurus. A respectable 20 feet long but only 11 feet high, it walked on all fours. But the hind legs were half again as long as the front ones, giving stegosaurus a hangdog posture. And most striking, from the back of its tiny head to the beginning of its tail was a double row of triangular horny spikes.

The precise purpose and arrangement of stegosaurus' armor is yet to be determined. The armor plates were embedded in the flesh along the spine and it is possible that the dinosaur could raise or lower them at will, either to make himself look bigger or to discourage the likes of allosaurus from biting its backbone. Perhaps the plates could be lowered to protect the flanks. Once stegosaurus was attacked, it arrayed its armor plate for defense, then lashed out with that whipping tail. On the end of it were four pairs of sharp, bony spikes, which could penetrate even allosaurus' tough hide.

In those distant times the great dinosaurs roamed virtually all the face of the earth. Their fossils have been found in North and South America, in Africa, Australia, Europe, India, China and Mongolia. All during the millions of years when man was developing from some apelike ancestor, these remains lay undiscovered, waiting, like dancers in an unseen back row, for the spotlight to fall on them. Primitive man, if he came across them, must have dismissed the fossils as one more natural phenomenon in the mysterious world he was laboriously beginning to conquer. Ancient man (with a few exceptions among the Greek philosophers) tended to regard big fossil remains as mythical giants who had once lived on earth.

The first documented traces of dinosaurs turned up in the United States but were not recognized. About 1800 one Pliny Moody lived up to a first name honoring a great Roman scientist when he found fossil footprints of dinosaurs in the Connecticut Valley. He had no idea what they were, but their birdlike, three-toed shape led some observers, who believed that Biblical animals were of huge size, to opine that they must be the footprints of the raven Noah sent from the ark in search of dry land.

William Clark, of the Lewis and Clark expedition, undoubtedly found dinosaur bones in 1806 below Billings, Montana. He understood no better than Pliny Moody the nature of his discovery, but both his description and his spelling are beguiling: "I employed my self in getting pieces of the rib of a fish which was Semented within the face of the rock this

William Buckland

Richard Owen

rib is (about 3) inches in Secumpherence about the middle it is 3 feet in length."

Remains of the first dinosaur to be identified and described as such were discovered in March 1822 in Sussex, England, by sharp-eyed Mary Anne Mantell, wife of a physician who was fascinated by fossils. One day Mrs. Mantell picked up a rock that looked as if it had a tooth embedded in it. When her husband saw it, he rushed excitedly back to the site for more and eventually sent a collection of teeth and some bone fragments to Paris to the greatest expert of the period—Baron Georges Cuvier. Cuvier identified the teeth as those of an extinct rhinoceros and the bones as those of an extinct hippopotamus. Dr. Mantell was not satisfied with this explanation, however. In 1825, quite by accident, he ran into a man who had been studying the large iguana lizard of Mexico and Central America. Dr. Mantell showed him one of the strange teeth, and both men decided it looked very like an iguana tooth, but larger. Thereupon Dr. Mantell published a description of his fragmentary fossil and gave it the name *Iguanodon* (iguana tooth). Baron Cuvier later graciously acknowledged his own error and foretold that a whole new group of animals—which he did not name —would be discovered from such fossil remains.

At about the same time, Dean William Buckland, a cleric and Oxford professor, studied some strange bones and a lower jaw that had been found near Oxford and decided that they had belonged to a big, meat-eating reptile he named megalosaurus. By 1842 so many big reptilian bones had turned up that Richard Owen proposed to the British Association for the Advancement of Science the recognition of "a dis-

DINOSAUR DISCOVERERS

Two English paleontologists (shown in the portraits, left) and the French anatomist at right helped establish dinosaurs as an extinct group of reptilian monsters. Baron Georges Cuvier of Paris, who pioneered the classification of fossils, was consulted by William Buckland of Christ Church, Oxford, England, about some large fossilized bones found near Oxford. Buckland then published a paper in 1824 describing a 40-foot-long elephantine reptile. Almost 20 years later Richard Owen of Richmond Park gave the reptiles a name: Dinosauria.

Baron Georges Cuvier

tinct tribe or suborder of Saurian Reptiles, for which I could propose the name of *Dinosauria.*"

Owen became so enthusiastic about the whole subject that he helped a sculptor named Waterhouse Hawkins construct a life-sized restoration of iguanodon; its completion was duly celebrated in London's Crystal Palace grounds with an elegant dinner on the last day of 1853—a dinner served inside the iguanodon model. The restorers, still not too certain how their monster had looked in life, placed on iguanodon's nose, rhinoceros-fashion, the big spike that the bipedal dinosaur had actually carried on its thumb. No matter. Iguanodon, flawed though it had become in resurrection, was a worldwide sensation.

Americans poring over newspaper accounts of iguanodon had no idea, yet, that their own continent would prove one of the world's richest in dinosaur fossils. The big United States dinosaur hunt did not get under way until just after the Civil War, when two eminent scientists named Othniel Charles Marsh and Edward Drinker Cope each determined to be the first on his block to have a dinosaur collection. Both were talented and wealthy. They loathed each other. Cope, later of the University of Pennsylvania, led an expedition in 1876 into Montana, where geologists had spotted fossil remains. He found the teeth and bones of a score of different dinosaur species.

Meanwhile, Marsh, a professor at Yale, was exploring in western Kansas and Colorado, and his crews had ventured into the now-famous happy hunting ground of dinosaurs, the Morrison geological formation in Colorado, Wyoming and Utah. The big dinosaurs that lived and died there during the Jurassic Period were preserved in river deposits laid down

At an 1853 banquet inside a reconstructed dinosaur, paleontologist Richard Owen toasts predecessor experts named on signs. The evening, said

the Illustrated London News, was filled with "philosophic mirth"—for which food and drink are readied at lower right, near a piece of dinosaur.

Edward Drinker Cope

RIVAL FOSSIL HUNTERS

Professors Edward Drinker Cope and Othniel Charles Marsh became bitter antagonists in the 19th Century rush to find and collect dinosaur bones. Cope wrote his first notes on fossils at age six and at 19 published a paper on salamanders under the aegis of The Philadelphia Academy of Natural Sciences. Marsh was born so poor that he could not even start high school until he was 21, but he went on to academic honors at Yale, where he was a cofounder of the Peabody Museum.

as the land slowly rose and the inland seas retreated. River-borne silt and debris covered the corpses and preserved them for millions of years until further upheavals of the earth made new highlands, and new erosion patterns exposed the fossils of the long-dead rulers of the West when it was really wild.

For more than two decades Marsh, Cope and their crews of hard-working diggers roamed the region with ever more sophisticated eyes, finding dinosaurs. They tried to work in utmost secrecy, to beat each other to the good finds. Their diggers once got into a fistfight in Wyoming, and the two professors brawled openly in newspapers and scientific publications. It was like a battle between brontosaurus and allosaurus, and the entire scholarly world rattled while it lasted. But when it was over, some of the world's richest troves of dinosaur bones had been found; selected United States museums and universities had enough material to keep experts busy for decades; and the way was open for a smooth transition from the age of swashbuckle to the age of cool assessment in the ongoing study of the terrible lizards.

Discoveries are still being made, however, and many are surprising. In 1964, for example, a whole new type of dinosaur, called *Deinonychus,* was identified in Montana. Deinonychus was fairly small, three and a half feet tall and eight feet long. It was a carnivore like big allosaurus and the even bigger tyrannosaurus rex. But the exploration team from Yale's Peabody Museum, led by John H. Ostrom, detected two strange features of the little dinosaur: First, it had a unique system of bundles of tendons enclosing the vertebrae of its tail; and second, it had a foot never before seen on a dinosaur. It had three

Ready for a dinosaur dig, Othniel C. Marsh (back row, center) and his students look more like Indian fighters than paleontologists.

toes, as do all of the carnivores, but the inner, or second, toe was armed with a long, thin claw.

Furthermore, the little dinosaur was obliged, by its very skeletal structure, to stand and move on two feet. To use its cutting blade it would have had to balance and leap on one foot while slashing with the other. Such action implies a high degree of dexterity and balance. Ostrom believes that the strange tendons in its tail were designed so that deinonychus could make its tail absolutely rigid. This ability made the tail what Ostrom calls "a dynamic stabilizer, and active counterbalance . . . like the tail of a cat." He followed dinosaur tradition in naming the new discovery: *Deinonychus* means "terrible claw."

That a new kind of dinosaur should suddenly turn up after a century of hunting is no shock to paleontologists, for during the terrible lizards' long reign they experimented with a dazzling array of forms, sizes and shapes. All are descended from some primitive reptiles that emerged during the late Paleozoic Era, about 240 million years ago, when for the first time the earth provided attractive edibles in the form of plants and insects. Some sea creatures responded to the new food by evolving so that they could live partially on land. Gradually some changed from fish forms into amphibians, others into reptiles.

The first in the main line of the reptile class were the lizardlike romeriids, no more than three or four inches long. Gradually they freed themselves from the water and their descendants acquired the ability to lay eggs on land. This innovation gave reptiles a strong advantage over the amphibians, which had to return to the water to produce their young, and it freed them to roam and increased their chances for both survival and biological evolution. Among the forms that appeared were the dinosaurs' most recognizable ancestors, the thecodonts. Over the passing millennia thecodonts grew ever longer and stronger legs until finally they were able to get up and outrun almost any other species then living. No sooner had they achieved relative superiority on earth than they began to evolve into a broad variety of other creatures. Some became crocodilelike; some became flying reptiles; and some became the two main types of dinosaurs, the reptile-hipped saurischians and the bird-hipped ornithischians.

Early in their history the saurischian dinosaurs split into two groups. One, the sauropods, included the largest of all dinosaurs but remained vegetarians. The other group began as smallish, meat-eating creatures that occasionally walked on their hind legs. Some of these then grew to tremendous size and became more and more bipedal. In the case of one of the best known of the meat-eating saurischians, the monster tyrannosaurus rex, the forelimbs had shrunk to almost useless appendages. Not that they were so small—each forelimb measured about three feet, for tyrannosaurus rex was more than 50 feet long, as tall as a two-story building, with a five-foot head, a four-foot jaw and six-inch teeth.

The bird-hipped, or ornithischian, dinosaurs were, to a large extent, even more varied in their evolutionary adaptations than were saurischians. All of them were plant eaters and some of them were bipedal, but they adapted their forelimbs as useful tools for both locomotion and food gathering.

Ornithischians went in for strange head shapes and a whole battery of bizarre armor. Among their forms were the duck-billed dinosaurs with broad, flat snouts admirably suited for shoveling up food from the muddy bottoms of streams and ponds. They also had webbing between their toes, just as any water-going creature should. The oddly armor-plated stegosaurus with the spiked tail was an ornithischian experiment, as was ankylosaurus, built low to the ground like a giant armadillo, with armor on its head and across its arched back, and a row of sharp spikes around the edges of the armor. Unlike stegosaurus, ankylosaurus had no spines on its tail, but the tail ended in a clublike mass of bone that as it swung must have created devastation behind the beast.

The last group of dinosaurs to appear, toward the end of the Mesozoic, was ceratopsia, the horned dinosaurs. Among them was protoceratops, which had a large head with a bony frill extending back from a turtlelike beak to a sweeping helmet over the neck. Horns, neck frills and heads all developed rapidly in this group, culminating in one of the most massive of dinosaurs, triceratops.

Triceratops stood eight feet high at the hips, was from 30 to 35 feet long and boasted one of the biggest heads of all the ancient reptiles. At the end of its nose was a short, stout horn and over the eyes were two long, pointed horns. Behind them grew a bony upswept frill. Triceratops' neck and leg muscles were enormous, giving it the capacity for short thundering charges and a lunge with those terrible horns. It must have been a match even for tyrannosaurus, particularly since one expert has estimated that triceratops could gallop at 30 miles per hour.

A few years ago anyone attributing such speed to a dinosaur would have been derided. Reptiles, the argument went, were cold-blooded and sluggish, with low metabolisms and small brains. The large plant

Text continued on page 84

A Gentle Collector Caught Up in a Dinosaur War

A self-portrait shows geologist Arthur Lakes drawing a fossil-bearing ridge.

In the late 19th Century, two kinds of battles raged in the Wild West: There was the famous sort between cavalrymen and Indians, and a less famous but almost equally bitter kind that pitted dinosaur collector O. C. Marsh of Yale against dinosaur collector E. D. Cope of the University of Pennsylvania. These two noted experts were hell-bent on beating each other to the fossils then being found in the West. And in 1877 they swept into their scholarly war an unassuming geologist, Arthur Lakes *(above)*, who had chanced upon the bones of a 60-foot fossil, the largest yet found. Lakes's role in the battle is told in his own watercolor sketches and writings, some of which are reproduced here.

The modest Lakes, hoping for assistance in his digging, had sent bones to Marsh, writing: *Whilst I am thoroughly embued with the enthusiasm attached to such pursuits and discoveries and should greatly like to continue them I have not the pecuniary means to do so.* Marsh, who already had a crew of fossil hunters on his payroll, ignored the plea—until he heard that Lakes had also sent some specimens to his archenemy, Cope. Immediately, he announced Lakes's find in a scientific journal, telegraphed his chief collector, Benjamin Mudge *(below)*, a geologist, to race to Lakes's

Workmen prepare to blast surface rock.

A skilled collector, Benjamin Mudge (right) checks Lakes's discovery at the open dig.

William Reed, Professor Marsh's Wyoming foreman, rides to a dig at Robber's Roost.

site in Colorado, and fired off reproaches and a belated check to the young fossil hunter.

Lakes replied: *Allow me to thank you for your generous assistance and the $100 enclosed. Funds were running very short; despairing of hearing from you, I was on the lookout to close with anyone who would help me.* Later: *when I forwarded those skulls to Prof. Cope I knew nothing of the reputation you give him.*

While money was Lakes's great need, he also valued the technical assistance Marsh sent. *In the afternoon as we were sitting at dinner under the trees, a gentleman rode into camp on horseback who turned out to be Prof. Mudge. I was very glad to meet him and to have someone of his experience and scientific knowledge as well as company and sympathy to aid me in the work. I took him over the ground and showed him what we had done. He seemed exceedingly delighted—and in amazement almost—at the very largest bones of dinosaurians or any other saurians he has ever seen.*

Lakes scraped by in Colorado on less than $50 a month for the next two years until the bones gave out in 1879. Marsh, adroitly manipulating from a distance, dispatched the ever-obliging geologist to a second fossil field, at Como Bluff, Wyoming.

Marsh's foreman at Como was one William Reed *(left)*, a gun-toting plainsman who had discovered the site and contracted to mine its bones. No one could fault Reed's loyalty to his master—at one point he destroyed several important fossils rather than let Cope's men get their hands on them—but he soon came into conflict with the scientifically minded Lakes. He heaped contempt on the young geologist's well-bred ways and his habit of recording strata and fossils in meticulous drawings (which turned out to be of immeasurable value to later generations of scientists).

"The Pleasures of Science" is Lakes's caption for this sketch of wintry work.

The lash of the environment was worse than the tongue of Reed. Gales and sandstorms swept through Como throughout the summer, alternating, Lakes wrote on August 9, 1879, with *heavy thunder and hailstorm stones like hen eggs.* Two days later he not-

While Reed supervises, hired hands tunnel into a narrow vein of ichthyosaurus fossils.

ed in his journal: *A heavy thunderstorm and rain occurred in afternoon in evening our tents were inundated with Siredons [lizards] who swarmed in such numbers insinuating themselves under every box and bed that although we threw out and killed dozens it became useless to stop the horde of slimy lizards that waddled leisurely into the tents as if they had a perfect right to them.*

Winter brought severe snowstorms that dumped 10-foot drifts on the digs *(left)*. When the temperature dropped to 38° below zero, beards, eyebrows and ears froze.

After 11 months of heroic labor at Como, Lakes finally left the service of the plundering professor and went on to earn academic laurels in his own right, teaching geology at the Colorado School of Mines. Yet he never escaped the gigantic saurians' thrall. In 1914, in his 70th year, he reconstructed the life of the dinosaurs in the paintings on this page, an old man's enchanting, if occasionally erroneous, reveries on a long-lost era.

Lakes in old age drew anatosaurus as he imagined it—but its tail was incorrect.

A trio of ceratopsians, shown with misshapen heads, browses by the Mesozoic sea.

Imagination afire, Lakes painted an allosaurus in mid-leap at a downed ceratosaurus.

An anchiceratops ogles a triceratops brazenly stripped down to its naked skeleton.

eaters, according to this view, must have spent most of their time half-submerged in water, to help sustain their enormous weight; they dragged their big tails along behind them until they could wade into water deep enough to float them. But recent and imaginative work by Yale's Ostrom, Robert T. Bakker of Harvard and Dale A. Russell of the Canadian National Museum, among others, is changing that picture rapidly and radically.

After a careful study of the anatomy and mechanics of the forelimbs of living vertebrates, Harvard's Bakker has concluded that the dinosaur had limbs mechanically almost identical to advanced mammalian forms. Instead of the awkwardly projecting forelimbs previously attributed to dinosaurs, there were, he now believes, limbs that could point straight down from a mammalian kind of shoulder joint.

The entire structure of the saurischian dinosaur's limbs and body, says Bakker, indicates a terrestrial animal like the elephant, with massive padded hind feet. It did go into water, perhaps quite often, but its feet would easily have mired in mud; like the elephant, it was well equipped to walk on land.

Even more interesting, Bakker thinks that dinosaurs may have had four-chambered hearts, like mammals and birds. This advanced type of pump efficiently moves blood through organs that remove wastes and replenish oxygen for the body's energy-producing processes, and thus sustains a high level of physical activity. Of modern reptiles only the crocodile has a four-chambered heart.

Some experts reject the idea of the four-chambered heart but agree that the dinosaur must have had some way to keep its body at relatively level temperatures —another mammalian ability crucial to an active life *(Chapter 4)*. They join Bakker in concluding that the dinosaur was certainly not a slow, sluggish thing.

Scientists are also taking a new look at the dinosaur's brain—the butt of jokes since the lovable beasts were discovered. When the first fossil remains of stegosaurus were found in the 19th Century, startled paleontologists noted that the brain cavity was only about the size of a Ping-Pong ball, and that a swelling of the spinal cord, in the lumbar region, was actually 20 times the size of the brainpan. This fact led one scholar to conclude that stegosaurus had two brains, and it inspired a Chicago wit named Bert Leston Taylor to a bit of doggerel. Stegosaurus had "two sets of brains," wrote Taylor: "One in his head (the usual place),/The other at his spinal base./Thus he could reason 'A priori'/As well as 'A posteriori.' "

In point of fact any vertebrate with arms and legs has one bulge of nerve cells in the spinal cord at the upper end to handle signals for the arms and a second, at the lower end, to carry messages to legs and tail. Neither is really a brain. But without the crucial lower communications center, Princeton Professor Glenn L. Jepsen has pointed out, perhaps as much as two seconds might elapse before a nerve impulse could travel from the tip of a big dinosaur's long tail to its brain and then back again with orders for action. If the creature's tail were being nipped at the time by a hungry predator, says Jepsen, "a lot could happen in a whole thirtieth of a minute."

Although dinosaurs were well equipped with centers for nerve messages, they nonetheless possessed very small brains. An iguanodon, about the size of an elephant, had a brain one twentieth the size of an elephant's, and the more diminutive dinosaurs fared no better. Modern mammals are without doubt smarter than dinosaurs were. But the dinosaurs were superb for their time, much more intelligent and active than anyone formerly gave them credit for.

The re-examination of past ideas about dinosaurs has inspired reinterpretation of such familiar old traces as the dinosaur trackways in Texas, where the footprints of dinosaurs, made 100 million years ago, have been preserved in mud that turned to rock. One trackway, along the Paluxy River, shows the enormous steps of a plant eater that walked along in shallow water while being pursued by a smaller, bird-footed carnivorous dinosaur. Each huge hind footprint of the intended victim is big enough to hold 15 gallons of water, yet the depth and evenness of the impressions dispels the idea that the creature had to walk in fairly deep water to support its weight. Whatever made that trail by the Texas river was moving in very shallow water indeed, with its legs well under it and its tail well up.

Other trackways, at Bandera, Texas, have preserved the trails left by at least 23 half-grown and mature dinosaurs. If they traveled in groups like this, they too must surely have held their long tails up to keep them from being trodden upon by their fellows. Movement in groups, as the Bandera footprints indicate, is far more typical of mammalian behavior than of reptilian behavior. Further, the tracks indicate that smaller, perhaps younger, individuals marched in the middle, protected on all sides by older or larger individuals—much in the way herds of elephants are known to travel today.

Further evidence of nonreptilian gregariousness in dinosaurs may lie in Mongolia, where paleontologists found a field of nests full of protoceratops eggs plus remains of what seem to be more than 100 individuals *(pages 94-95)*. The presence of so many eggs in one place suggests group egg-laying to some scholars. Whether or not the females usually stayed to hatch their eggs is moot: reptiles normally do not, but at the most important Mongolian egg site there were remains of mothers, babies and eggs. There is also evidence of a sudden sandstorm, which could have overwhelmed them all without warning.

Perhaps the most convincing, if least specific, argument of all for the biological superiority of dinosaurs is the very length of time during which they dominated the earth. Their survival depended, as does that of all organisms, on a food chain beginning with sun-nourished plant life. Yet the earth's flora changed notably during the 155 million years of the Mesozoic, and dinosaurs adapted and flourished.

They avoided extremes of temperature, yet apparently they lived in such diverse regions as deserts in Mongolia, open plains in Africa and forests in Europe. In North America they survived inundation that eliminated much of their living space. They were top dog on earth for 135 million years, about 133 million years longer than man has achieved.

What, then, killed the kings? Their era ended, abruptly by geological time, at the end of the Mesozoic, and it ended all over the world at about the same time (though some forms of dinosaurs had died out earlier). No dinosaur remains have been found in the deposits of the Paleocene, which followed the Mesozoic Era. Geological evidence indicates that whatever happened affected all forms of life, not just dinosaurs. It killed about half the species of flowering plants, many primitive varieties of mammals, the flying reptiles and the big swimming reptiles.

To explain this disaster, some far-fetched theories

A combat to the death between two dinosaurs—a scene often imagined but never documented—was finally revealed 100 million years after it took place with the uncovering of these two skeletons. Frozen in struggle, a meat eater called velociraptor (black) grasps between his forelimbs the head of his prey, the plant-eating protoceratops (red). At this moment both creatures suddenly died together, no one knows how, and were caught forever in their unfinished fight. They remained buried until 1971, when a combined Polish-Mongolian expedition found the two small dinosaurs—neither is longer than seven feet—in the sandstone of the Gobi.

have been offered over the years: a catastrophic plague that infected all the great reptiles; a sudden taste for dinosaur eggs on the part of the emerging mammals; sterility caused by climate changes; a form of racial senility in which an entire family, like one individual, simply got old and passed on.

Most logical of all the explanations is that a drastic, though perhaps brief, change of climate killed the dinosaurs off. They could not tolerate wide fluctuations of temperature such as those that prevail in most temperate zones today. If the earth suddenly got quite cold, the dinosaurs (and many other forms of life) would have died. But the chill could not have lasted long, for there is no evidence of an ice age.

What could have caused such an abrupt cold snap? Suspicion of some stunning cosmic event, such as a sudden burst of radiation, has been offered tentatively, and this idea gains credence with recent research. Canada's dinosaur expert Dale Russell and Wallace Tucker of the Cambridge, Massachusetts, firm of American Science and Engineering, Inc., believe that the extinction of the dinosaurs is specifically related to the astronomical phenomenon of supernovas.

Over the past 4,000 years, men have seen and reported at least seven times the sudden appearance of a dazzlingly bright star, visible both day and night, which then gradually faded away. These transiently bright stars, the supernovas, are now considered the result of tremendous stellar explosions, in which the density and heat of the core of a massive star finally exceeds its limits of tolerance and the star blows up, releasing huge amounts of various kinds of energetic radiation—cosmic rays, gamma rays and X-rays, and of course, visible light. None of the recorded supernovas blew up within 100 light-years of the solar system—near enough, that is, for its radiation to alter the environment on earth. A close supernova is likely only about every 50 million years, too rare an event to be expected within the span of human existence but a quite probable one over the longer span of the dinosaurs. In 1971 Dale Russell described the twofold effect of such a nearby explosion. First, the earth would receive a heavy and deadly shower of gamma rays at ground level. Second, the blast of X-rays emanating from the same explosion would blow off a portion of the earth's atmosphere and the X-rays' enormous energy would quickly be deposited in the form of heat in a layer extending from 12.5 to 50 miles above the earth.

"The resulting turbulence," Russell explained, "would probably disrupt the heat-retaining properties of the atmosphere, generate many storms of hurricane force at the earth's surface and circulate low, water-laden air into higher, drier levels. There it would freeze to form a high-altitude cover of ice clouds, which would reflect much of the sun's heat away from the planet. The net effect . . . would be to cause surface temperatures to drop all over the world and severely tax or exterminate organisms adapted to tropical climatic conditions."

This hypothesis is, as Russell points out, based on incomplete evidence. But it does explain the relatively sudden extinction, and it fits with geological evidence of a dramatic but brief climate alteration.

It also fits, eerily, with the general upgrading of the evolutionary achievements of man's favorite "failure": the largest and most highly evolved terrestrial organisms, the big flowering plants and the dinosaurs themselves, would have been worst hit both by the increased radiation and by the sudden cold.

The Bellicose Life Style of the Dinosaurs

Asked to imagine what life among the dinosaurs was like, most people conjure up a gory fight scene like the one at right. They are not far wrong. The giant reptiles did indeed spend much of their time battling one another.

Sometimes the saurian quarrel was among meat eaters, each bent on enjoying a freshly killed meal. Sometimes it was a dispute over territorial rights. Sometimes two males clashed over the issue of sexual dominance. Sometimes, as at right, it was a fight to the death between a hungry meat eater and its prey. The picture shows the carnivorous deinonychus, only eight feet long and 175 pounds in weight, attacking the much larger tenontosaurus. Although the latter was a vegetarian, it was hardly a milquetoast. It stretched to 25 feet in length, weighed a ton and had powerful hind claws that could punish an aggressor cruelly. But deinonychus had the advantage of great agility plus an extremely well-developed balance apparatus, which allowed it to keep its place on tenontosaurus' back while slashing away with its lethal weapons, a sharp, sicklelike claw attached to each hind foot.

Despite such ferocious battles, dinosaurs ordinarily got along well enough with one another. Within species they engaged in many communal activities, nesting in groups and traveling in herds that seem to have been organized for their common defense.

In search of a meal, a meat-eating deinonychus attacks a plant eater, tenontosaurus.

Massive Meals to Feed Hungry Giants

Grazing ravenously are, top to bottom, camarasaurus, stegosaurus and camptosaurus.

Mealtime among the dinosaurs must have been a sight, for the big ones were more than just big eaters; they were probably the most voracious land creatures that ever lived. About three fourths of a ton of leaves and twigs was the daily ration for the camarasaurus at left, which weighed four times as much as an elephant. Huge size helped account for such giant appetites, but there were also other reasons: Most large dinosaurs ate plants, a less efficient source of energy than meat, and they were not sluggish—as most scientists recently thought—but were active animals that burned up food at a great rate.

Their massive hunger was easily satisfied by the rich savannas and forests of the Mesozoic Era, for the earth then had a generally mild climate, and many areas now barren were lushly green. There were giant ferns crowded around lake shores. Dotting the savannas were strange trees with leaves rather like modern palms, some only a foot or two tall, others, like the williamsonia in the drawing at left, reaching 35 feet.

The bountiful savannas became communal mess halls where each dinosaur ate at the level it could most easily reach. And if there was enough food for plant eaters, that meant plenty of food for meat eaters as well. Carnivores like tarbosaurus and deinonychus were always near, ready to pick off a stray *(right)*.

While two tarbosaurus giants wrangle over a kill—an 18-foot euoplocephalus—it is seized by a six-foot cousin of deinonychus.

A Ritual Battle, Head to Head, over Sex

With backbones and tails held horizontal for running, two dome-headed dinosaurs crash together in a mating-season test of strength.

There was once a dinosaur known as *Stegoceras* but more descriptively dubbed domehead. Domehead's skull was roofed with bone three to four inches thick. This feature gives one of the few clues to the sex habits of these ancient creatures. The thick skull existed apparently because male domeheads would square off in pairs, lower their heads and charge into each other. The domehead that butted the hardest presumably won a harem of female domeheads.

Scientists who advance this theory base it on the similar actions of modern mountain rams, which will butt heads to establish mastery. Since there is evidence that the domeheads lived in herds, the domes may also have been symbols of rank in the herd, like the large horns of today's sheep that help establish dominance within their herds. The fact that the domeheads lacked spikes suggests that the domes were used only in ritual mating contests—they would have been too weak for lethal weapons.

The Communal Act of Egg Laying

Having selected a warm and sunny spot, two female protoceratops join in burying all their freshly laid eggs under a blanket of sand.

A nest of protoceratops eggs discovered in 1923 in Outer Mongolia's Gobi Desert revealed that this dinosaur, and presumably others, followed a social pattern rarely found in the animal world: they got together to lay their eggs in a communal nest. The evidence is the number of eggs the nest seemed to have contained originally —perhaps 30 or more, too many for only one protoceratops to have laid.

Each female apparently laid its clutch of eight-inch eggs in a pattern of ever-widening concentric circles. An inner circle had a few eggs, the next circle a few more and an outer circle a larger number. The fat end of each egg was tilted up. The surface at this end was smooth, in contrast to the wrinkled ridges that covered the rest of the eggshell. After the eggs were laid, the dinosaurs covered the group nest with sand, leaving the eggs to incubate by themselves. The embryos reached a length of 10 to 12 inches before they broke out—eventually to grow to six or eight feet.

Baby protoceratops emerge from the shells they have just pecked open. At this age they lack the neck frills seen on their mothers.

A Herd in Formation for Defense

When overgrazing diminishes their food supply, or if drought, flood or predators threaten their existence, modern mammals that live on the African plains migrate in herds to greener pastures. Dinosaurs may have done the same, according to a recently revived theory. The evidence lies in patterns formed by fossil footprints, which indicate that groups of dinosaurs moved along a straight path at a steady pace toward some destination that now lies beneath countless strata of earth. Exactly what the animals were leaving behind and what they sought at the end of the trail,are still mysteries, but the large-scale migrations seem to have been impelled by the need for fresh sources of food.

These dinosaur herds apparently were organized in a way that shows a most unreptilian concern for the young. Today's reptiles generally ignore their offspring, aside from eating one occasionally. But the tracks left by one group of brontosaurs indicate that the largest animals walked on the periphery of the column while the smaller ones stayed in the center. The plausible explanation is that the full-grown males were guarding the herd to shelter the youngsters from meat-eating marauders.

A group of brontosaurs in formation—powerful male lookouts positioned around weaker juveniles in the middle—crosses a plain.

Chapter Four: In Man's Body, Debts to His Past

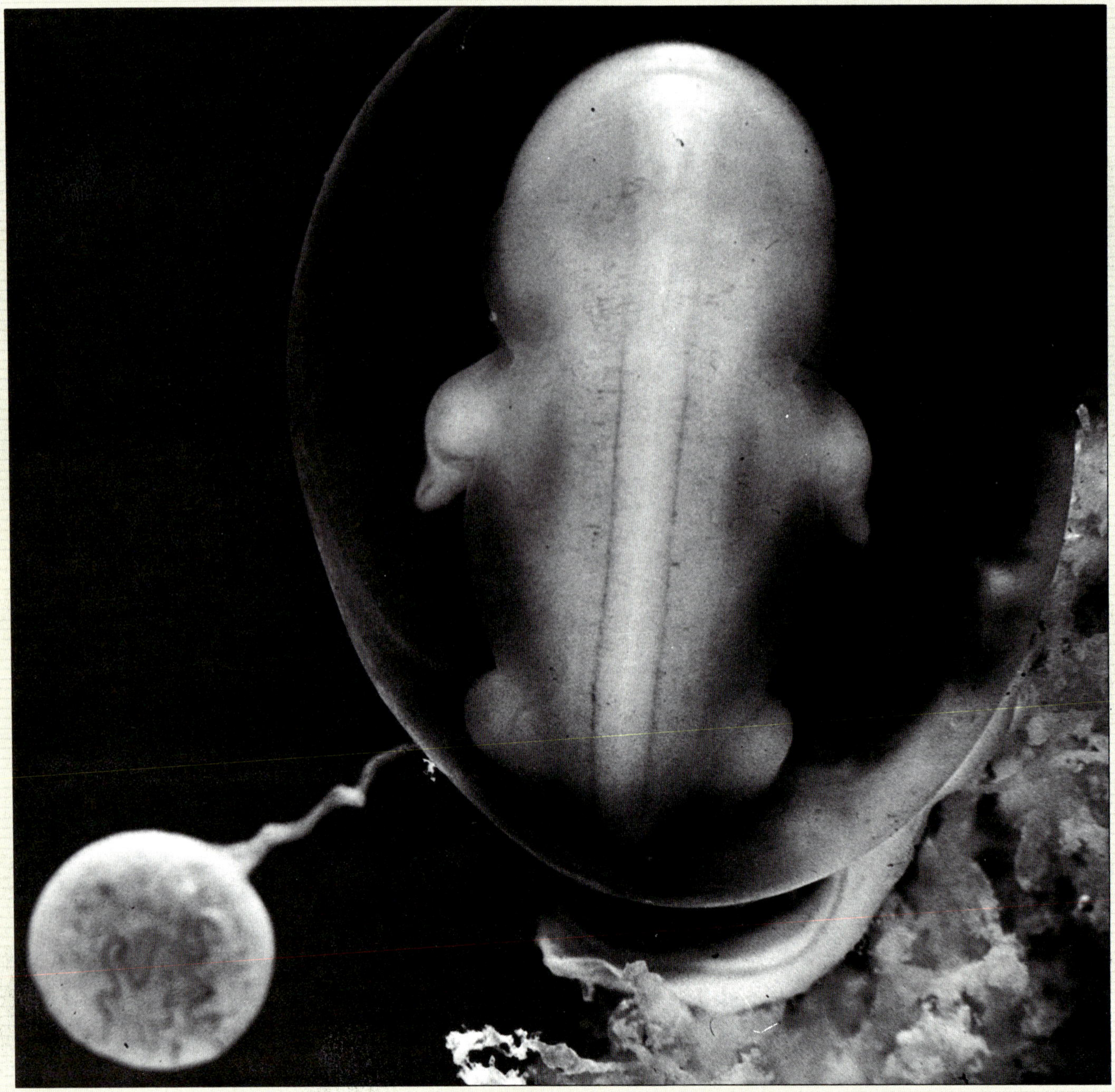

"Evolution," wrote the eminent geneticist Theodosius Dobzhansky, "is a synthesis of determinism and chance, and this synthesis makes it a creative process. Any creative process involves, however, a risk of failure, which in biological evolution means extinction. On the other hand, creativity makes possible striking successes and discoveries."

The dinosaurs were a spectacular failure. After a reign of 130 million years they vanished from the earth, leaving no descendants. But another experiment in the creation of animal life, starting from the same ancestral sources as the dinosaurs, did not end in an evolutionary blind alley. Down through the ages it preserved a thread of life that, despite many strange twistings and turnings, ultimately led to what Dobzhansky aptly describes as "the greatest success of biological evolution to date"—modern man.

The long and remarkable history of the way man acquired the attributes that make him uniquely successful has left the human body full of traces of ancestors that were very different from man and led wholly different lives. A few of these relics are useless or nearly so. At the end of the human spine is a coccyx, the vestige of a tail that various furry animals must have found useful as a blanket or balancing organ. Certain other vestiges such as the appendix can give a good deal of trouble. There are also a few ways in which the body has not developed evenly. The heads of human infants, for instance, have grown so large to contain the all-important brain that they barely pass through the female pelvic opening, making childbirth often difficult. But all in all, the human body is amazingly efficient, the product of trial and selection that began more than three billion years ago.

A six-week-old human embryo, protected by its mother's amniotic sac, floats in saline fluid as man's oxygen-breathing aquatic ancestors did nearly a billion years ago. At this stage of its growth the embryo resembles a fish more than a human, with its flipperlike arms and legs, its spine ending in a tail and gill-like pouches that will soon become a lower jaw.

During this enormous span of time nature alone influenced the developments that eventually provided the human body with its internal skeletal support, its constantly warm temperature, its legs for walking upright, its hands for deft manipulation and the other significant features that would enable man to dominate the earth more completely than even the dinosaurs had done. The evolution of all animals was a response to the challenge of the environment. If the climate was cold, natural selection favored the development of fur and fat. When supplies of grass and leaves spread, so did crunching teeth. Every animal was at the mercy of its surroundings; if it suited its environment it prospered, and if it did not, it either changed to meet the conditions in which it found itself, or moved somewhere more suitable, or died. This dominance of natural environment over evolution came to an abrupt end once man appeared upon the scene. If he had to find food by preying on other creatures, man did not need to develop fangs and claws; he made weapons of wood and stone. When the climate turned cold, he wrapped himself in animal skins instead of growing a furry coat.

For a million years or more, man's evolution has been independent of his surroundings, and his adaptability to any environment—even the hostile vacuum of space—seems assured. But today he is ready to tip the balance between evolution and environment in another way. He is now able to interfere directly with processes established by his own evolution; in recent years he has acquired, often without realizing what he was doing, the ability to change the genetic inheritance that makes humans human. Such power can lead to disaster: a drug that seemed a valuable remedy for sleeplessness had tragic effects when taken by expectant mothers, so distorting the normal development of their unborn babies that limbs resembled primitive flippers instead of arms and legs.

And yet the precise knowledge of inherited mechanisms that is now accumulating promises the possibility of controlling some of man's genetic weaknesses. Certain burdens that humans carry in their evolutionary inheritance might be lifted—mental retardation could be prevented, for example, if its genetic instructions could be erased. Perhaps it would even be possible to stimulate another great step in evolution such as further expansion in the capacity of the brain. But how successfully man will use his emerging power to steer the course of the future may depend on how well he understands the steps by which nature formed him in the first place.

The basic pattern of the body that we take so for granted—an interior and exterior, a front and rear, a right side and a mirror-image left side—got its start in the warm seas that rolled over nearly all of the ancient earth. The body's most fundamental feature is simply an inside (distinguishable from an outside) alimentary canal to handle food and water. It originated among the very early inhabitants of the primeval seas—the first multicelled animals, hardly more than specks of jelly. One of these primitive creatures eventually acquired an interior tube into which food materials were drawn, were exposed to processes of digestion and then were expelled at the other end. This device proved to have so many advantages that it is now standard equipment not only for man and all the other higher animals but for many of the lower ones as well.

The alimentary canal was a great invention partly because it is a canal, open at both ends, and partly because it is inside the body. Since the cells lining the tube are protected by their interior position, they can be delicate, thin-walled and therefore more efficient absorbers of nutrients than if they were on the outside. But in addition, the straight-through design keeps wastes from getting mixed up with the incoming food, as they often do in creatures whose digestive tract is a simple sac with only one opening. The flow of food material is easily controlled, and digestive juices released in the confined space of the tube act more efficiently than they would outside. In higher animals the canal is equipped with elaborate valves, holding tanks and pumps, but its function has not changed in nearly a billion years or more, nor has its vital importance diminished. Even the highest animals, including man, can be described as mechanisms whose life depends largely upon how they supply and protect their alimentary canals.

The primitive multicelled organism that first acquired an alimentary canal automatically acquired a front and a rear, the front being the end where the food enters and the rear being the end where wastes are expelled. If such a fore-and-aft animal can crawl or swim, it moves in the direction of its front end, seeking food for its hungry tube. If it has organs of sight and smell to lead it to food, and tentacles or teeth to capture it, these naturally cluster around the tube's entrance, which then can be called a mouth. And when the creature's nervous system becomes elaborate enough to need something resembling a brain to sort out its signals, the best place for this control center is near the concentration of sense organs, where its services are in greatest demand. Very humble animals have this close grouping of mouth, brain and sense organs, and so does man.

From remote sea-dwelling ancestors also comes another of the basic features of man's structure: his bilaterally symmetrical body plan. For man and most living animals that walk, swim or fly freely, a body having one side identical with the other has proved to be the most successful. It gives the ability to move front end first and steer from side to side while keeping right side up. Only the body's exterior need be symmetrical; the interior parts not concerned with locomotion may vary in shape, although many of them are central or paired. Animals that move slowly or not at all are apt to have partial symmetry like the snail, or radial symmetry like the five-armed starfish.

The establishment of the body's fundamental pattern was a first step in the general direction of man, and for about three billion years the creatures that exploited its advantages ruled in the early seas. They were all invertebrates—many of them soft-bodied, boneless things, like modern squid; others, like lobsters and insects, wore a bonelike structure outside their bodies. But with the appearance of fish, the first animals with interior spinal columns, the invertebrates' age of glory was over. The road of evolution had taken its most crucial fork, and from this point on, the vertebrates—of which man is the supreme example—gradually rose to dominance over the insects, crustaceans and other invertebrates, which have always been far more numerous.

The importance of the spinal column cannot be overemphasized. It provides the foundation around which man's internal skeleton is built. Every animal that moves vigorously benefits from some stiff material to which to attach its muscles. For arthropods such as insects and crustaceans, the exterior skeleton provides protection and attachment points for muscles. A serious disadvantage is that the animal cannot grow bigger without shedding its skeleton and secreting an entire new one. This process is not only costly in bodily material, it is also extremely dangerous. For instance, a lobster that has just shed its shell, or molted, is utterly defenseless and must hide under rocks or seaweed until its new shell has grown hard. Partly to avoid this perilous period, lobsters and most other arthropods remain small, so that frequent shedding is unnecessary. Many insects never shed their external skeletons and do not grow once they pass the skeletonless pupal stage.

It is the internal skeleton that permits vertebrates to grow large while still remaining active and efficient. They have no dangerous molt to worry about. The bones inside their flesh are not discarded periodically but continue to grow in harmony with the rest of the body.

The first animal to have such an invaluable structural form was probably an early fishlike one resembling the modern lancelet, or amphioxus, a creature that lives in shallow sea water and looks like a small, translucent minnow. It is, however, a great deal simpler than a minnow. It has no jaws, no teeth, no paired fins, no bones. It does not pursue prey as minnows do. It is a filter-feeder like a clam, most of the time burying itself in sand or mud with its forward end protruding to take in water and food;

The Evolving Spine

From the simple structure in a prehistoric fish to a complex instrument in modern man, the spine has evolved to support body and head and to aid intricate movements.

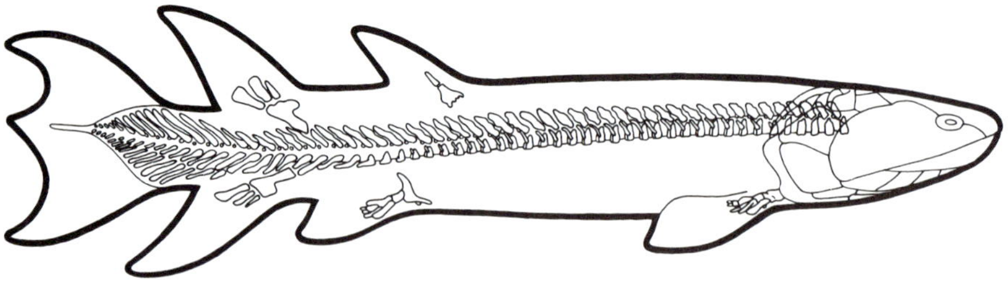

An undifferentiated spinal column served eusthenopteron, an early bony fish of 375 million years ago. Its similarly shaped vertebrae, joined to short ribs, gave swimming muscles something to pull against.

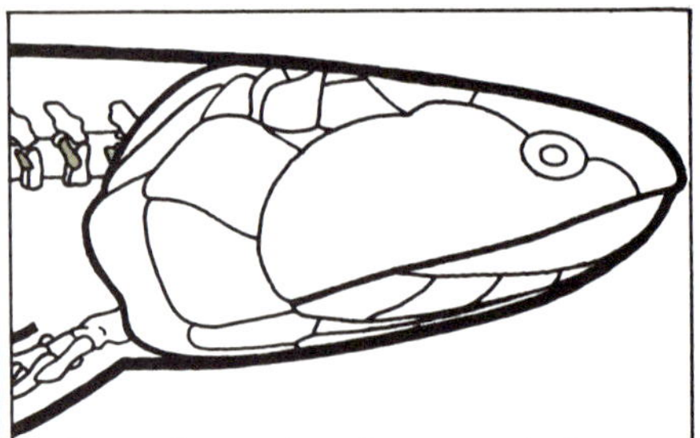

The uniform ribs (olive green) along eusthenopteron's spine lent only a lateral undulating movement.

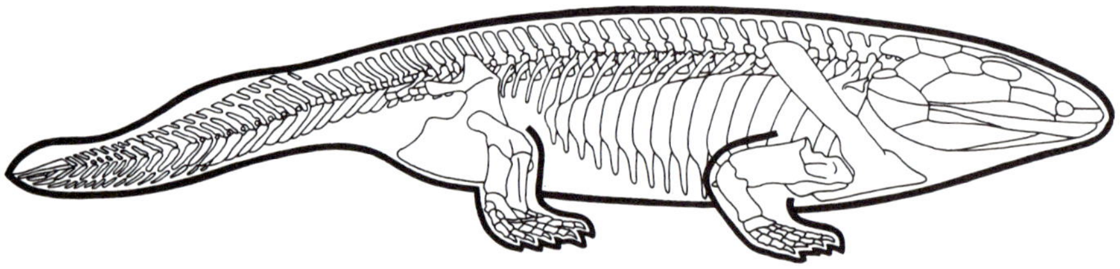

The amphibian ichthyostega required a sturdier spine than eusthenopteron because on the land there was no water buoyancy to help support its body weight. Its vertebrae, as a consequence, were more solidly constructed.

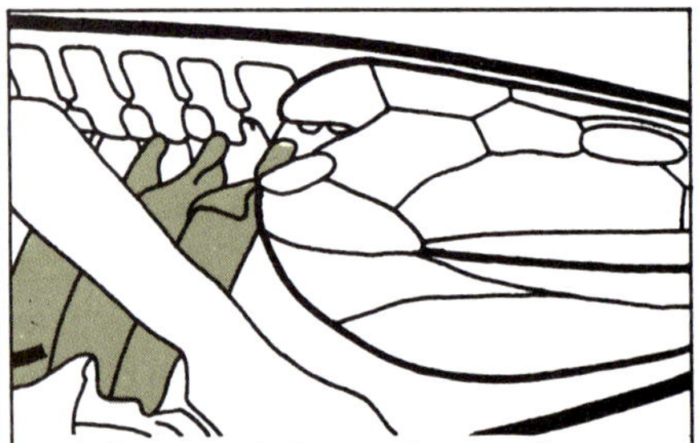

Ichthyostega's large ribs may have helped it to hold up its head on land as well as supporting its body.

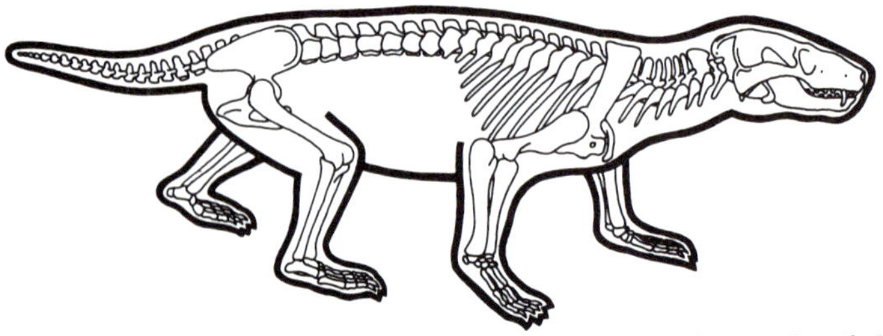

The vertebrae of the mammallike reptile thrinaxodon, even more closely locked together than those of ichthyostega, had specialized shapes and sizes: for example, large near the limbs, smaller in the lighter tail.

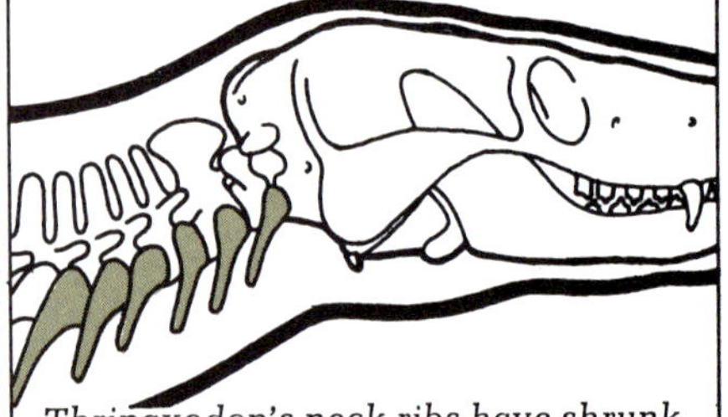

Thrinaxodon's neck ribs have shrunk, enabling it to move its head far more easily than ichthyostega.

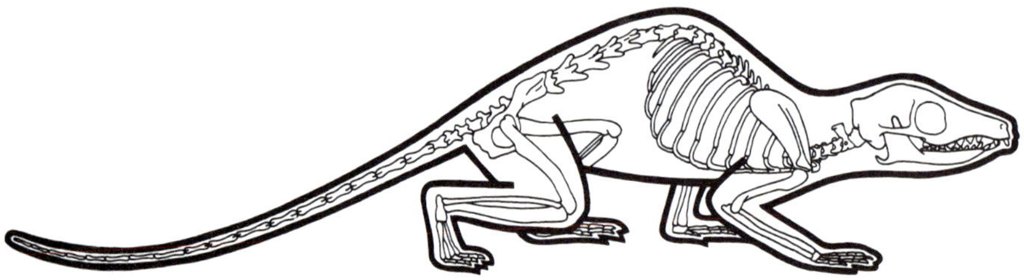

A modern tree shrew that resembles extinct primitive mammals moves along the ground as well as climbing trees, arching and extending its backbone as it goes. Its vertebrae are designed for both types of movement.

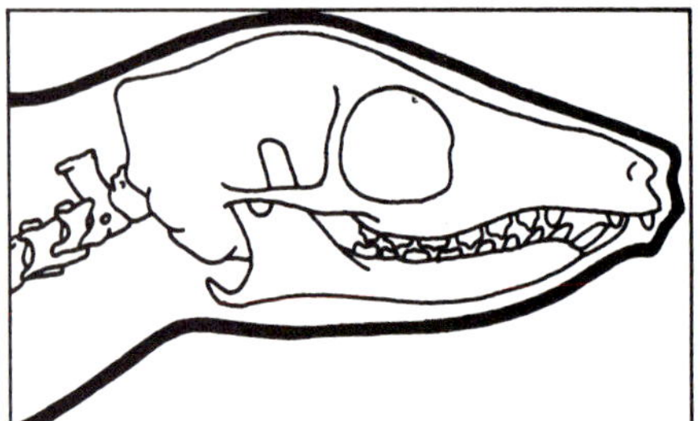

The tree shrew's highly flexible neck results partly from the shrinkage of neck ribs, now mere vestigial nubbins.

Although a quadruped, the ancient primate mesopithecus was capable of briefly supporting its body on its rear legs while reaching and grasping, and its backbone was accordingly specialized—rigid when upright but flexible enough to allow it to travel through trees. Its vertebrae have acquired a variety of shapes. The small cervical, or neck, vertebrae permitted head movement while supporting the skull either vertically or horizontally. The large vertebrae in the lumbar region of the lower back supported propulsive movement.

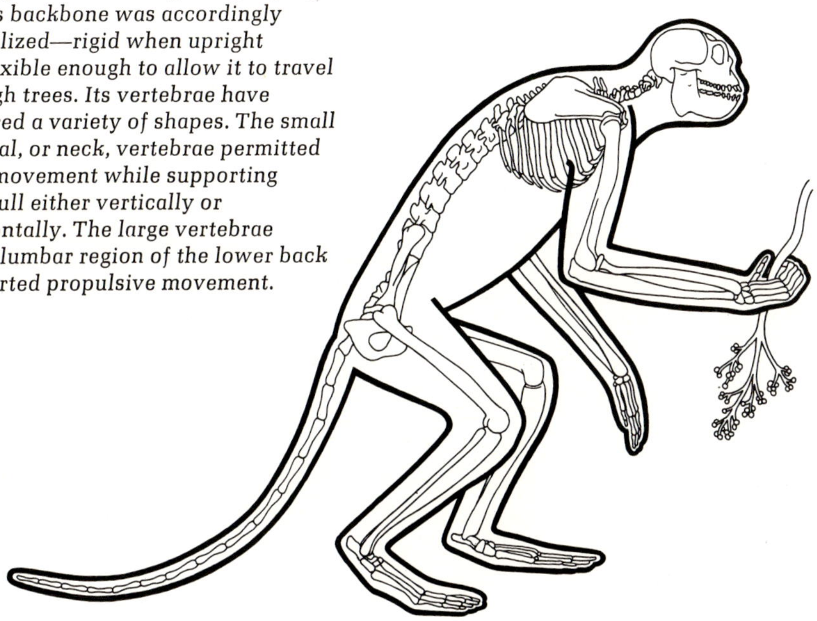

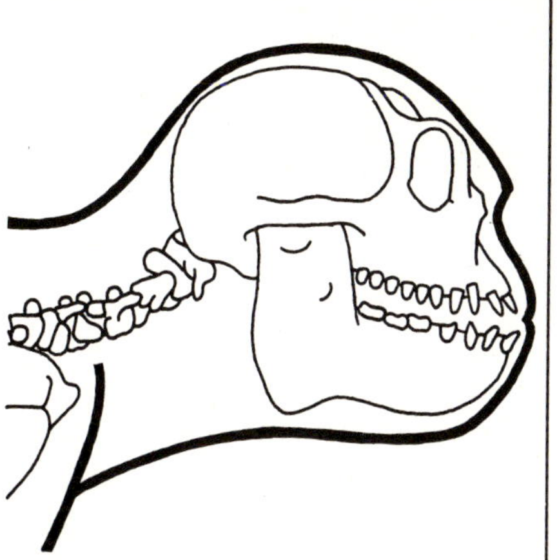

Mesopithecus' head movement depended partly on the "atlas-axis complex" of two neck vertebrae. The top one, called the atlas because it supported the skull in somewhat the way the Greek god is supposed to have held the earth, enabled the head to move up and down—and thus is known as the "yes" bone. The axis, just below, helped give sideways head movement—hence, it is the "no" bone.

To provide support for man's upright, bipedal posture, the vertebrae of his spine are strongly locked together in a flexible, vertical rod. The vertebrae are increasingly heavy from the top down to the hip, where the weight of the body is transmitted to the legs. The backbone must not only be strong enough to bear most of man's weight, it must also be flexible enough so that he can balance on two legs. The compromise is not always entirely successful: man's vertebrae are separated by easily damaged discs, and back trouble is a common complaint.

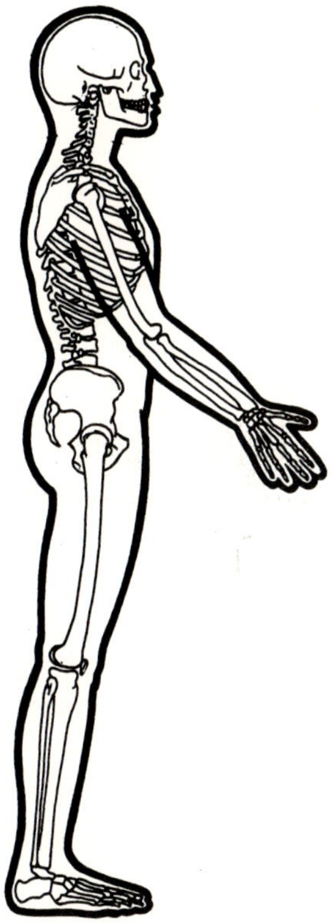

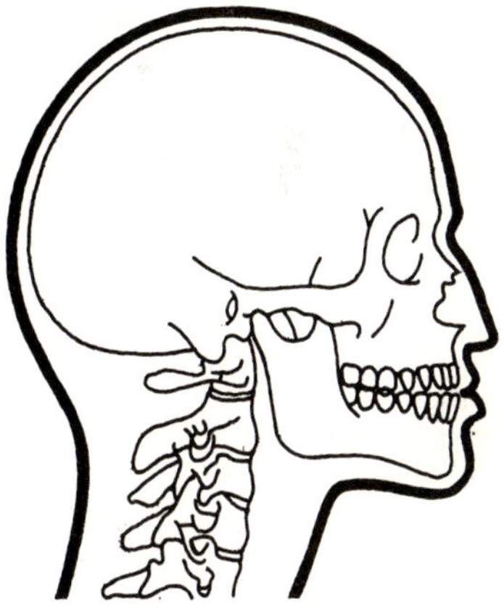

Man's upright posture has also given him a head position that in relation to his spine is different from the position of the heads of semi-erect primates. The top of the spine has migrated from its position in back of the skull, in mesopithecus, to a point almost directly under the skull. Thus man's head is neatly balanced at the top of his fully erect spine, and there it stays as he freely moves his ribless neck.

the food sticks in its mouth while the water quickly passes out through slits in the sides of its body.

This way of life is not typical of an active, mobile vertebrate. But within its sluggish body the lancelet does have features of great evolutionary significance. Among them is a bundle of nerves, equivalent to the human spinal cord, that runs down its back and is slightly enlarged at the forward end like a rudimentary brain. Just below it is a fibrous rodlike structure that allows the lancelet to bend its body sideways but prevents telescoping. This is the organ around which the spinal column developed millions of years ago. Its technical name is the notochord.

Man owes to the fish not only the bones of his backbone but other bones that link up closely with it. They look to be outgrowths of the spine—jaws, teeth and skull. Yet they did not develop from interior bones of any primitive creature but, in the strange ways of evolution, from external features of early fish.

The skull may have come first. In man, as in all higher animals, the skull is a case of sturdy bones fitting right at the end of the spine as though one developed from the other. But originally some of these bones were plates of external armor protecting the primitive brain of an acanthodianlike fish. In early fishes those plates became covered with skin to form the internal structure of the head.

In the course of evolution, the notochord of some presumably lanceletlike creature became surrounded by a jointed series of bony segments that strengthened it and eventually replaced it. The original purpose of this more elaborate piece of apparatus was to enable fish to swim more efficiently. Fish swim by means of large muscles covering each side of their bodies. When the muscles contract in sequence, they bend the body into a series of waves that push against the water, and with the help of the tail fin, which flaps back and forth, these motions propel the fish forward. The great advantage of the backbone is that it gives the swimming muscles something to pull against and lets the fish bend its body into waves without compressing or distorting it, which would reduce the efficiency of the swimming action. Backbones made fish much better swimmers than they would have been otherwise and enabled them eventually to dominate the oceans.

When fish appeared in great numbers in the Silurian Period they seem to have lived in fresh-water ponds and streams where they sucked nutritious mud from the bottom through jawless, toothless mouths. Such mouths were useless against the large water scorpions and other predators of that age. Before fish themselves could become predators of anything besides very small creatures, they needed jaws and teeth that could slash and bite. As it happens, they had on either side of their throats a series of paired skeletal bars shaped like "V's" with the points directed backward. The purpose of these bars was to support the gills that the fish used for breathing and possibly for straining food out of the water. As the fish evolved, the first two pairs of bars apparently disappeared, but what was probably the third got bigger, acquired a hinge at the point of the "V" and became flexibly jointed, turning into bony jaws foreshadowing those of man and other higher vertebrates.

Jaws are generally not much use unless they are armed with teeth. Surprisingly, these did not develop as might be expected from the bony jaws of the ancient fish but from sharp points, or denticles, that studded their skin. (Sharks still have denticles, and they make shark skin feel like sandpaper.) As in sharks, the denticles in the skin along the edges of the newly developed jawbones of primitive fish were composed mainly of dentine, the principal tooth-forming substance. Apparently these points grew bigger and bigger until they became true teeth, effective as weapons, food catchers and food crushers. Much later, with the rise of reptiles with mammalian traits in the Permian Period, the dentary, or tooth-bearing, bone of the jaw became by far the largest of the seven bones of the typical reptilian jaw and began to suggest the single curved bone that makes up the lower jaw in man and the other mammals.

Much more obvious than the origin of skull, teeth and jaws is man's indebtedness to the fins of the early fresh-water fish for his landgoing limbs. Anyone looking at a fish can imagine it upright, waddling on its two rear fins and waving its front fins as rudimentary arms, like the anthropomorphized characters in animated cartoons. From the fins of such an upright fish it might seem only a few easy steps to a man's legs and arms—some bones have to be lengthened and a couple of jointed connections added.

Such a transformation may work in the movies, but it did not happen that way in the real world. If a primitive fish could have been stood on end, its fin bones (like those of a modern fish) would have pointed sideways at such an angle they could not hold the body up—and the place where toenails were eventually to evolve would have pointed to the back, not the front. Before fish fins could become human arms, legs and feet, some of the most remarkable alterations in evolutionary history had to take place (*drawing, page 106*). The short, relatively broad and rigid fin bones lengthened and narrowed, multiplied and became jointed. Platelike bones of the pelvis and shoulder developed to provide bases for muscles and fulcrums between limbs and spine. But, strangest of all, those bony structures that would become human toes ultimately had to change direction; by the time man had evolved, they had moved around almost 90° from their original orientation, so that feet pointed forward and lay under the body to support its weight for walking, while arms swung freely at the side, able to reach into almost any position.

The process of evolving limbs began some 400 million years ago. At the beginning of the Devonian Period, the Age of Fishes, some fish had acquired two pairs of fleshy, movable fins to control their swimming. In most cases, these fins were balancing aids only. They could not be used for efficient paddling. The rear fins ended in a rather small bony plate that was not attached to the fish's spine; without firm support, the rear fins had nothing to push against and therefore could not exert appreciable swimming force. The front fins, on the other hand, were firmly attached. But their connection was fairly rigid so that only limited movement was possible.

But among these early fish was a group known as crossopterygians, or lobe-finned fish; their fins developed stronger muscles and a set of jointed internal bones. On these sturdy fins the lobe-finned fish could crawl slowly out of the mud on their fin-legs in time of drought and make their way along the bed of a dwindling stream in search of a pool that still had water in it. An almost perfectly preserved fossil fish found in Pennsylvania in 1971 clearly illustrates the beginning of the branching of the bones that would ultimately form the upper arm, forearm, wrist and hand bones of the human arm.

From Fins to Limbs

The priceless human skills of walking on two feet and manipulating with two hands are the inheritance of changes that transformed fin bones to meet new demands.

The early precursors of legs and arms, such as the fins of eusthenopteron, helped primitive fishes to balance but were not much good for propulsion. Each rear fin (near right) was attached to a pelvis (olive green), but the pelvic bone was not attached to the spine, leaving the fins with no firm support to push against. Each front fin was joined to a shoulder girdle (far right) that was attached to the relatively immovable skull.

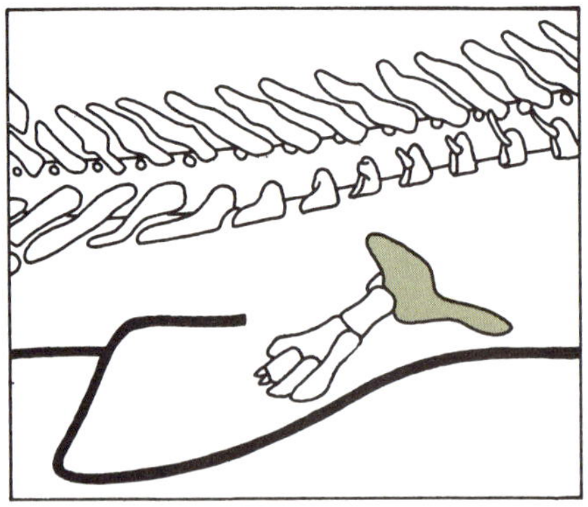

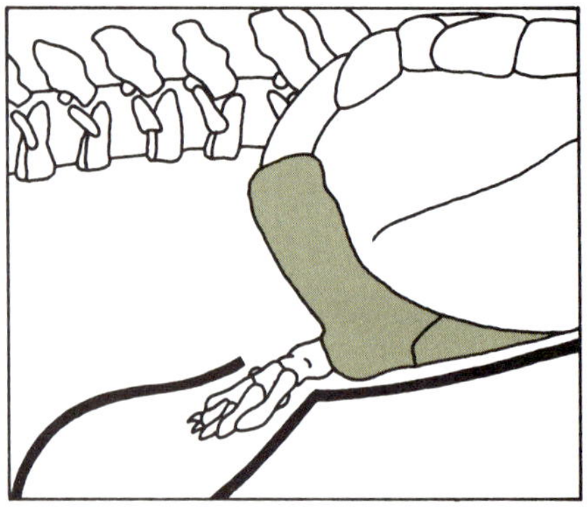

In the early amphibian ichthyostega the onetime fins swung to the side and developed joints to become flipperlike legs for waddling. The hind leg was then attached by a pelvic girdle to the spine, which provided the brace that allowed for forward movement. The shoulder girdle that held the front legs then became separated from the skull, providing greater mobility.

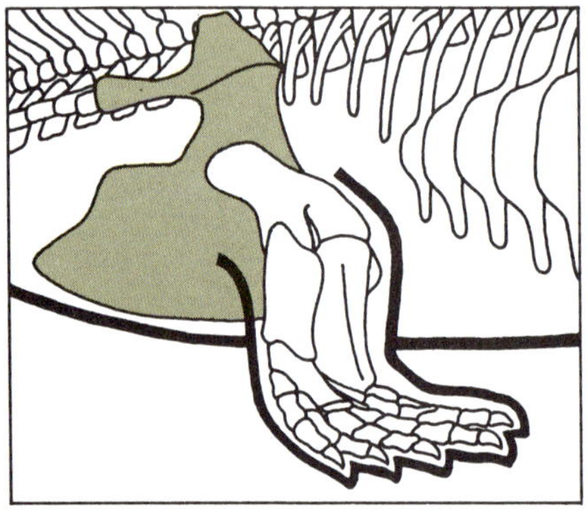

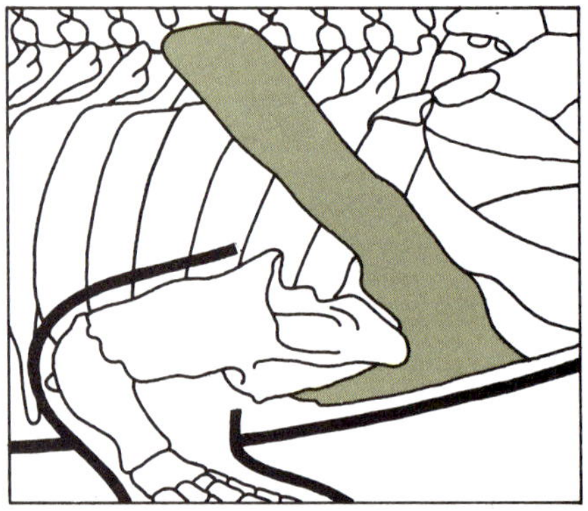

Thrinaxodon, a mammallike reptile, walked with agility. The broad attachment of the pelvic girdle to the spine gave added foundation for a powerful hind limb. The shoulder girdle became lighter, to increase the mobility of the foreleg. And both pairs of legs shifted from the side position of the amphibian to a mammallike location nearly under the body.

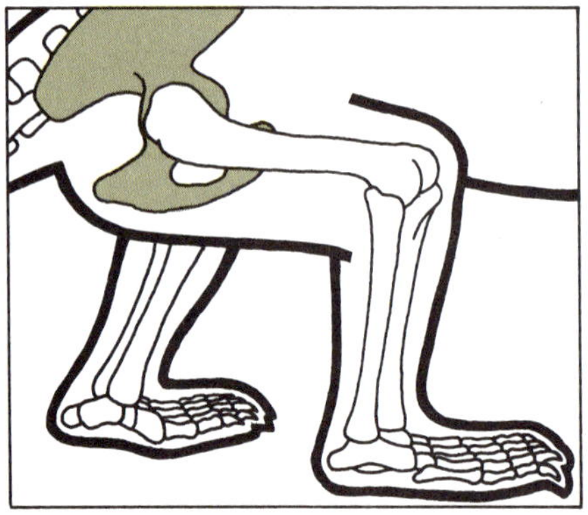

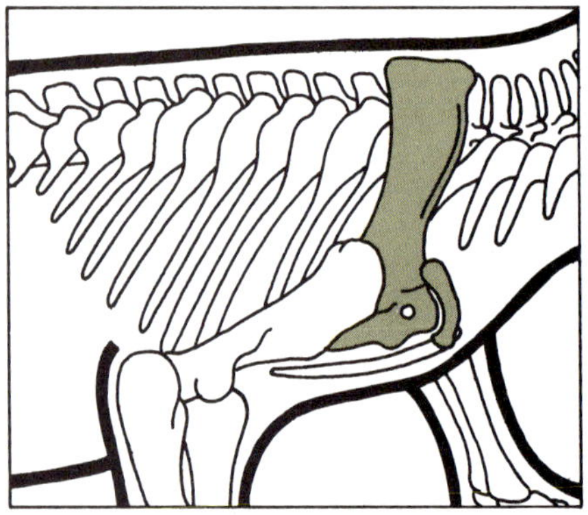

The pelvic bones of the modern tree shrew are elongated and narrow, accommodating muscle arrangements that fit the hind legs for agile arboreal and terrestrial life. A collarbone attaches the shoulder girdle (both in olive green at the far right) to the breastbone at only one point. This flexible joint enables the animal to rotate the shoulder and raise the arm.

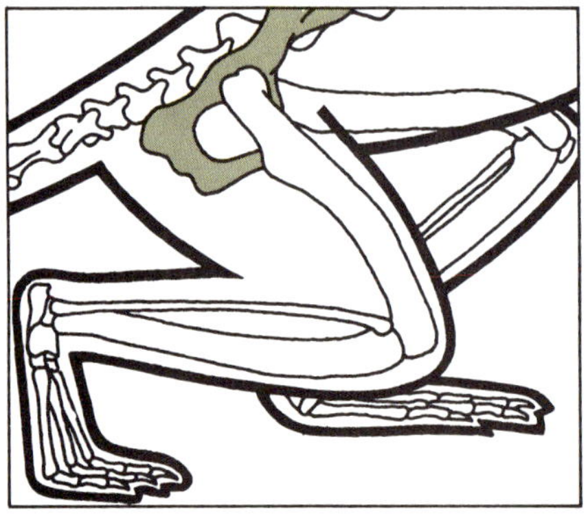

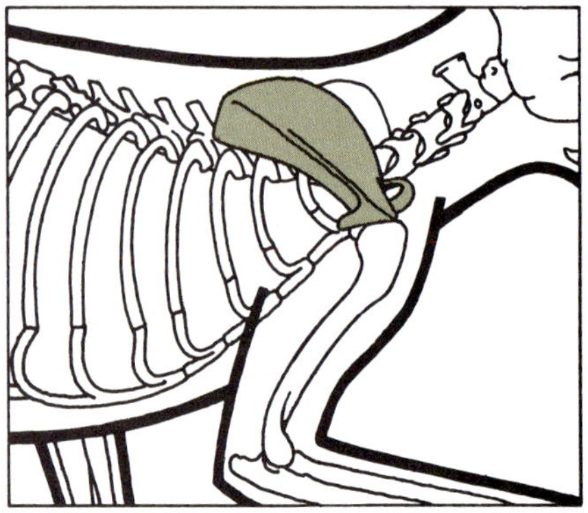

The early primate mesopithecus had legs well suited to quadrupedal walking—long and placed almost directly beneath the trunk. The pelvic bone extended forward along the spine and the forelegs gained mobility apparently because of the arrangement of the muscles that were attached to the spine and also because the shoulder joints were improved.

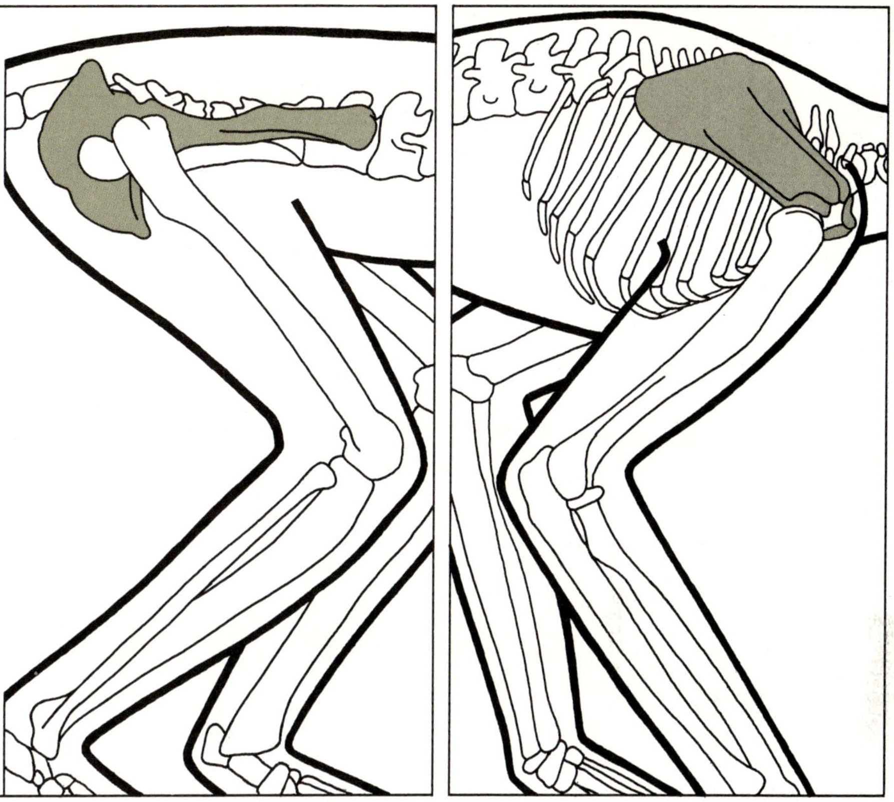

Man's uniquely effective scheme of walking on two legs depends on specialized bone structure. The pelvis consists of two parts, shown here in front (top) and side views, that are fixed to the base of the spine and transmit the weight of the trunk to the legs. The pelvis is short and wide, providing an extensive base for the hip and leg muscles. The human arm can be moved to almost any position since its end fits in a socket in a broad shoulder blade that can also move.

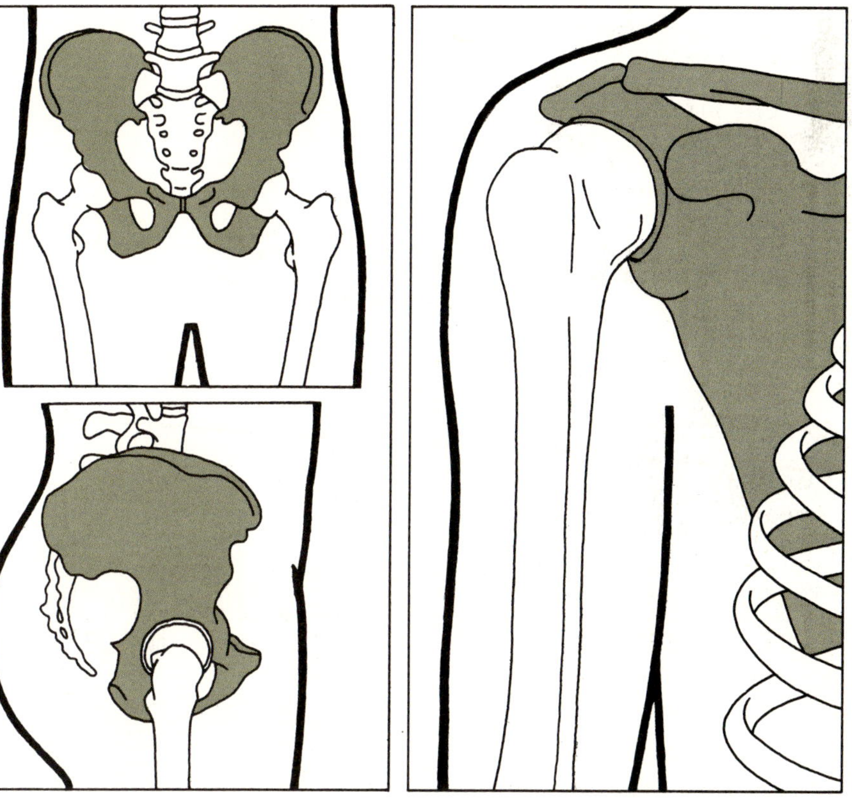

From these adaptable fish, at home in water or on land, arose the amphibians. The earliest were still very fishlike; they had the remnants of a fish's tail, useful for swimming, but what had been fins had become stubby, wide-spraddled legs for walking—if what the primitive amphibians did can be called walking. Their gait, like that of present-day salamanders, was a waddle because all four legs projected sideways. They were not underneath the body, where they would carry the animal's weight efficiently, nor could they move very far at each step.

But even the limbs of some early amphibians had five sets of articulated bones forming the toes. Toes are essential in running, and later vertebrates experimented with various numbers of them. For example, eohippus, the earliest horse, had four toes in front and three in the rear. Later forms of the animal lost toes; a modern horse has only one functional toe on each foot. It moves on the tips of these toes—elongating its legs to provide a very light foot and a long lever action for high-speed running. Man also elongates his legs for speed by going up on his toes to run, but he runs less than he walks so he normally uses all of his foot. He is extremely conservative in this matter; he still has the five toes with which early fishlike amphibians crawled out of the mud.

When the earliest reptiles made their appearance 350 million years ago, they still had wide-spraddled legs similar to those of their amphibian ancestors and walked with a clumsy, crawling gait. And most of the remnants of the great reptile class that survive today—such as crocodiles and lizards—are almost equally clumsy. But 225 million years ago, at the end of the Permian Period, there was a group of reptiles that apparently died out after giving rise to the earliest mammals; these mammallike reptiles were well equipped for walking and even for fairly fast running. Their limbs had made another shift in direction, moving from the amphibians' sideways position to a location more nearly underneath the body. In addition, the bones of the thigh in the rear and the upper "arm" in the front could swivel more freely at both ends, so that the legs operated quite differently from the outward-protruding amphibian limbs, which moved in an arc. The legs of these mammallike reptiles could move backward and forward parallel with the length of the animal's body, taking long, efficient strides. Also, their feet were turned so that they pointed forward; they thus could roll flexibly over the ground at each step, adding spring to the gait and providing stabler support during movement.

The reptiles were the first full-time land dwellers. They had efficient lungs for getting oxygen from air into the bloodstream. Reptile lungs were a great improvement over the crude air sacs that had appeared among some early fish apparently simultaneously with gills, and they were considerably better than amphibian lungs, which often had to be supplemented by a system that used the skin to absorb oxygen from water. But the reptile's great contribution was a scheme of reproduction that worked on land.

The amphibians, like their fish predecessors, had to breed in water. The females extruded eggs into water, and the males fertilized them there, apart from the females. From this point on, the eggs were generally on their own, to grow and develop into larvae that would eventually find nourishment in the water independent of their parents. The reptile's system was crucially different. The egg was fertilized inside the female's body. As the embryo within it developed, the embryo was surrounded by a fluid-filled sac called the amnion; provided with a container of yolk, for food, and a waste-disposal sac, the allantois; and wrapped in a tough shell. Only then, a ready-to-grow package complete with its own source of nourishment and protection, was the egg expelled to develop outside the animal.

The amniote egg remains the basis for human reproduction; although important changes differentiate the mammalian system from the reptilian one, there are more similarities than might be thought. An egg cell is still fertilized inside the mother's body. As it develops it is surrounded with an amnion, a fluid-filled sac like that in the reptilian egg.

The slightly salty fluid bathes the embryo, protecting it from shocks, and acts as an internal pond remarkably like the home of the fish ancestral to both man and reptiles. There is, of course, no shell for a human egg. Outside the amnion there is still a yolk sac, but it contains practically no yolk. Instead, the embryo is supplied with nourishment and relieved of wastes by the mother's bloodstream. This diffusion of vital materials takes place through the placenta, which may have evolved from the waste-disposing allantois of the reptilian egg. The nourishment received through the placenta enables the human embryo to develop inside its mother's body, where it is much safer than in a shell hatching in a nest.

Through the amniote egg, the reptiles bridged the transition of life from water to land, adapting to an environment in which mammals would evolve and reproduce their kind. The reptiles, in particular those whose skeletal structure shows limbs, jaws and other features approaching those of mammals, may also deserve credit for another key development on the road toward man—the beginnings of the temperature control that is to a large extent responsible for making man and nearly all his fellow mammals the most active and intelligent creatures on earth.

Man has elaborate systems to keep his body temperature constant within a few tenths of a degree, whereas the internal temperature of reptiles, amphibians and fish—in fact, all other animals except mammals and birds—fluctuates with that of their environment. The disadvantages of fluctuating body temperature are many. Lizards, for example, are torpid on cool mornings. Their bodies are so cool that the chemical reactions that animate their muscles proceed relatively slowly. The best the lizard can do to get its body functioning properly is to drag itself into a patch of morning sunlight. As the radiant heat warms its body and blood, its heart beats more effectively. Its muscles reach their peak of activity, and presently the lizard darts away at proper lizard speed. By contrast, man's activity is almost independent of the temperature of his surroundings. Within fairly broad limits he can run fast and work hard whether the day is cold or hot.

Not merely active life but life itself depends on internal temperature. All animals must keep the temperatures inside their bodies within a restricted range of their own; changes above or below that bring quick death, as anyone knows who has raised tropical fish in a living-room tank. The creatures most tolerant of extreme temperatures seem to be mosquito larvae—some types have been found in hot springs, where their bodies are at about 120°F., while others survive in Alaska even if ice forms within their tissues. The limits for man, whose blood temperature normally fluctuates only a fraction of a degree, are

Care of the Young

From amphibian to man, care given to the young becomes ever more complex, and the elaborate protection given a human baby indicates its evolutionary pre-eminence.

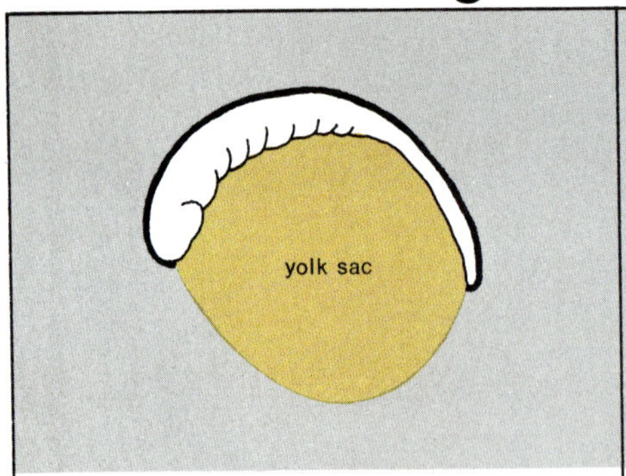

The 22-day embryo of the amphibious salamander necturus, or mud puppy, lies unprotected by a shell in the water the egg was laid in. It is fed from its large yolk sac (yellow).

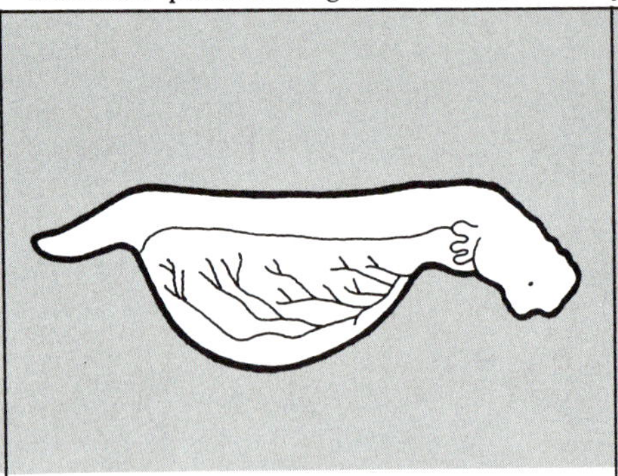

At 36 days, the yolk is almost all consumed, its material carried to the embryo by blood vessels. The mud puppy's head extends in front of the sac (right) and its tail behind.

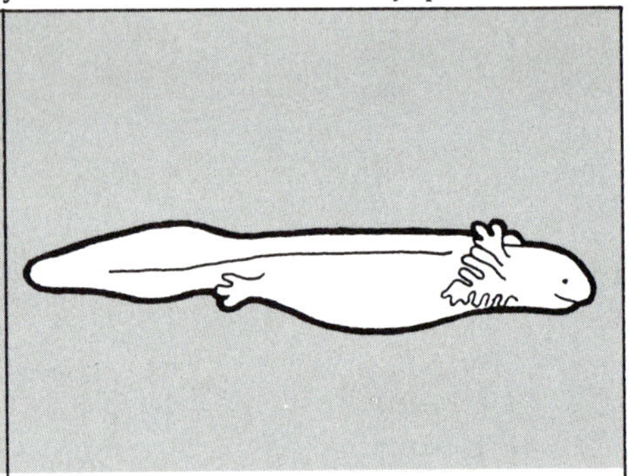

At 49 days, two weeks before incubation ends, eyes, gills and limb buds have developed. The yolk sac is consumed, and the salamander must now seek all its own food.

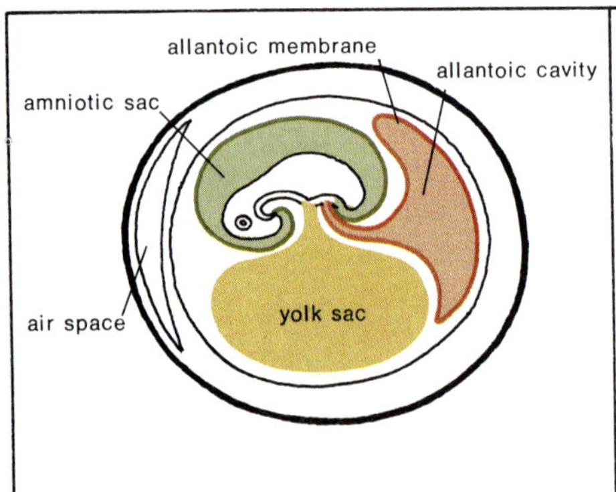

Reptiles develop in shell-protected amniotic eggs, named for the amnion (green), a shock-absorbing sac enclosing the embryo. Another sac, the allantois (tan), absorbs wastes.

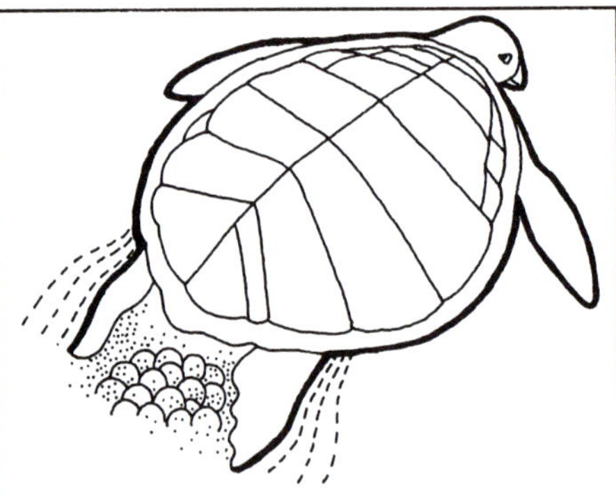

The green sea turtle lays its clutch of eggs on land in a hole it has dug. To reinforce protection of the eggs' shells, the turtle covers them with sand—but then leaves them.

Hatched after about 60 days, the baby turtles are miniature replicas of their parents. To avoid bird predators, they scuttle to the sea, where they must find their own food.

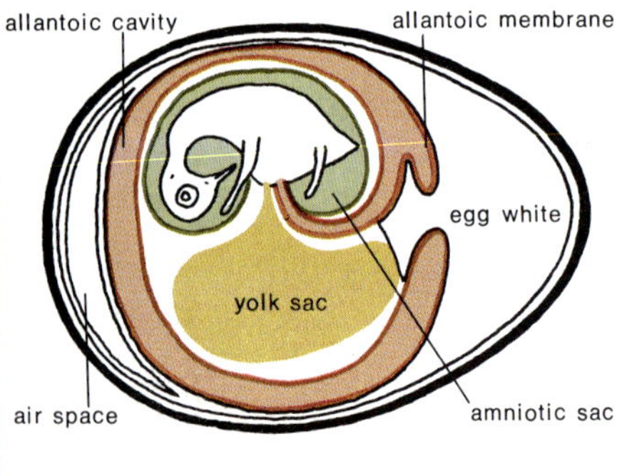

Bird eggs are similar to reptile eggs. But since this embryo is older than the turtle above, its allantoic sac is larger; supplied with blood vessels, it acts as an embryonic lung.

Bird eggs are not left alone to hatch but are incubated by the female, which sits on them for periods as long as 80 days. Her body heat keeps them at a constant temperature.

The young of most birds emerge from the egg blind, naked and helpless. They require parental care and feeding until they grow feathers and gain strength, usually in two weeks.

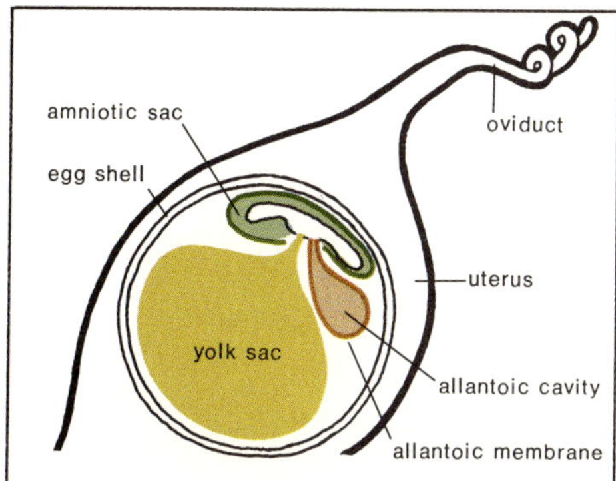

The embryo of the spiny anteater, an early mammal, lies in an egg in the mother's uterus. It is partly fed on uterine fluid that reaches the yolk sac through the egg shell (double line).

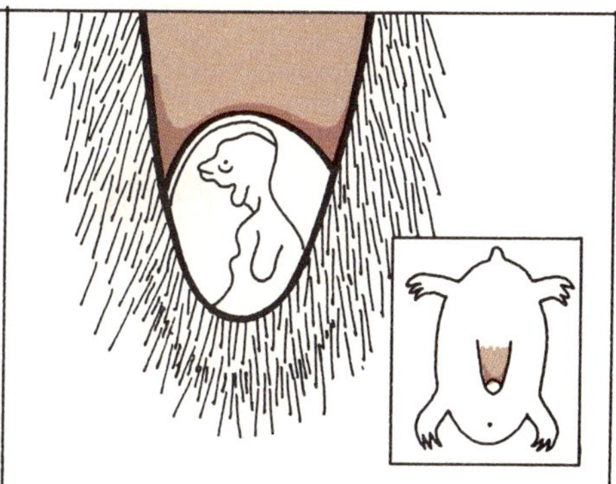

After laying an egg, the anteater somehow places it in a hairy pouch, where it incubates. Inset shows the position of the pouch in the center of the mother's lower abdomen.

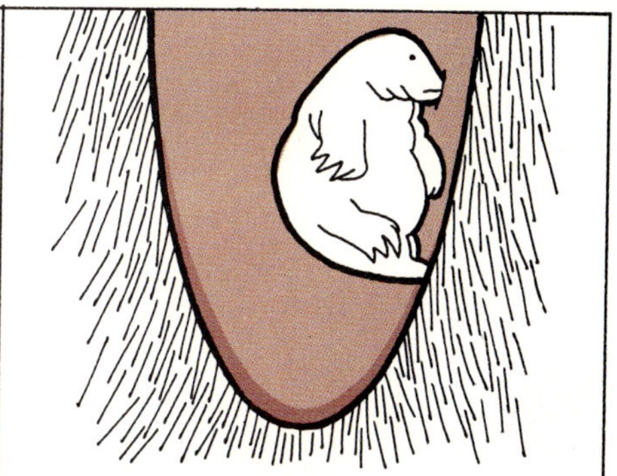

Hatched after two weeks, the infant gets milk from glands in the pouch lining. It stays in the pouch for 10 weeks, after which the mother cares for it until it is more fully grown.

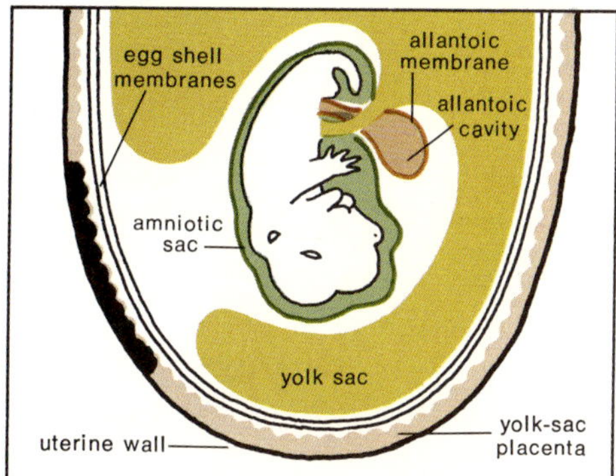

In the kangaroo the egg shell has become thinner. Nourishment is secreted by the uterine wall. It then flows through the egg shell layers to the yolk sac and on to the embryo.

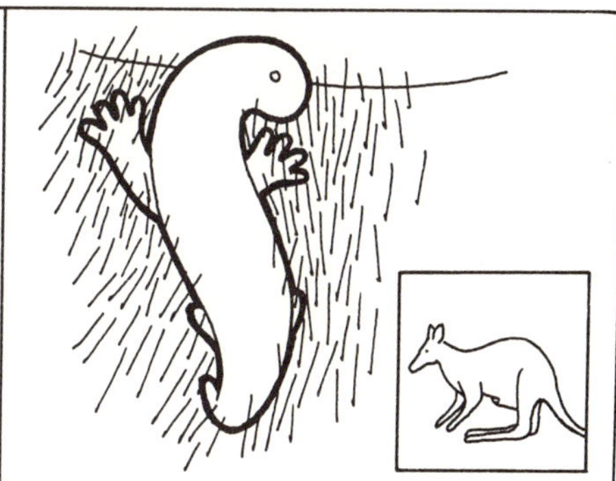

Less than an inch long, the kangaroo is born alive and returns to the warmth of its mother's body at once. Grabbing the hair on her belly, it climbs up into her pouch (inset).

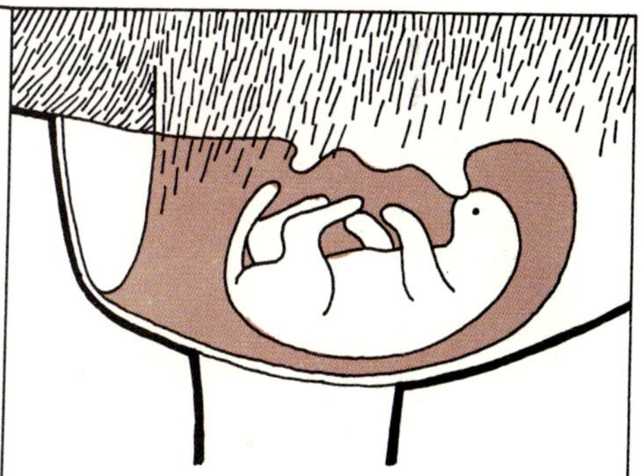

The kangaroo's mother has well-developed teats in her pouch on which the baby suckles. It leaves the pouch after four months but returns to suckle for many months more.

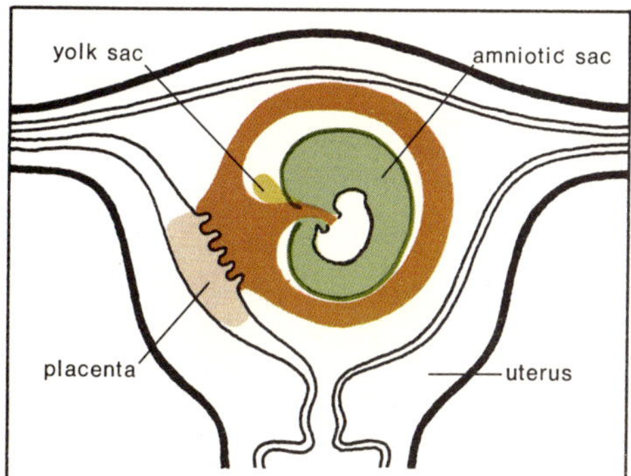

The human embryo is surrounded by tissue that contacts the uterine wall. This area, the placenta, removes waste materials and carries nutrients from the mother to the embryo.

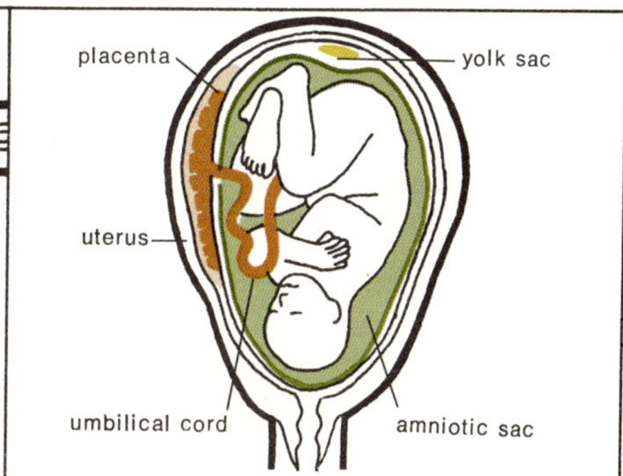

Because the human embryo depends on food supplied not by the egg but by its mother, its yolk sac is tiny and the fetus continues to be surrounded by a fluid-filled amnion.

At birth the baby can suck, which means it can feed at its mother's breast. It can also grip, cling, cry and cough. Otherwise helpless, it needs intense and prolonged maternal care.

very narrow; death usually comes above 109.4°F. or below 77°F.

The life-or-death importance of body temperature is related to its connection with physical activity. Activity requires energy, which is supplied by food-processing reactions within the body, and these metabolic reactions proceed slowly in the cold and rapidly in warmth. The rates of many animals double with an increase of about 20° in body temperature, but all biochemical processes do not change their rates in the same way. Thus, if the internal temperature goes very far over the normal upper limit, some scientists believe, one process may speed up so much that it produces more of an intermediary product than the body can handle; and conversely, if the temperature drops below the lower limit, a process may slow down so much it cannot produce the required amount of a product needed for another vital step. In either case, the body is thrown out of kilter by extreme changes in its own biochemical reactions.

Within the life-or-death limits of body temperature there is almost always one temperature at which bodily processes operate most efficiently. The development of ways to keep the body at that ideal point forms a principal theme of evolution. Fish in general have had less of a problem in this respect than land animals; the temperature of their watery habitat, especially large bodies of water like the sea, does not fluctuate nearly as much as that of other natural environments. In any ocean, it usually varies up to about 25°F. For example, Rockall Banks, a fish-rich area in the North Atlantic, has a winter temperature of 49°F. and a summer temperature of 56°F.

Land creatures, on the other hand, had to survive drastic alterations in air temperature—in parts of the American Midwest the thermometer drops below −40°F. in winter and in summer climbs over 110°. Reptiles and insects managed mainly by escaping the extremes—burrowing into the ground, hiding under rocks or retreating into water. Even so, their body temperatures went up and down, varying widely from the level of efficient energy production. But as evolution progressed, more advanced creatures smoothed out these peaks and valleys *(graphs of body temperatures, page 113)* until the birds and mammals came along, with their ability to keep body temperature constant at an ideal level.

Astonishingly, the best temperature is almost exactly the same for all mammals and birds. Man keeps his body at 98.6°F., a mouse at 97.7°F., a horse at 99.8°F. and an elephant at 97.1°F.; songbirds' bodies are about nine degrees warmer. Clearly, an internal temperature near 100°F. makes the processes of life operate at the highest safe rate (at higher temperatures many cells die).

To maintain the precise body temperature that permits his vigorous activity, man has acquired a whole battery of special equipment and actions to warm him up and cool him off as necessary. The evolution of these features is difficult—and in many cases impossible—to trace, because they involve nerves, blood and soft tissue that leave almost no trace in the fossil record. But the origins of some can be deduced.

One important need for temperature control is insulation. Man, like many animals, has a layer of fat under his skin—a blanket that evolved very early among the reptiles. Presumably he once had external insulation as well. He still has some hair on his body, and his close cousins among the apes have much more. Such fur is now believed to have appeared first

Controlling Heat

As heat control improved, animals kept body temperatures even, despite air temperature changes—increasing their activity and decreasing dependence on environment.

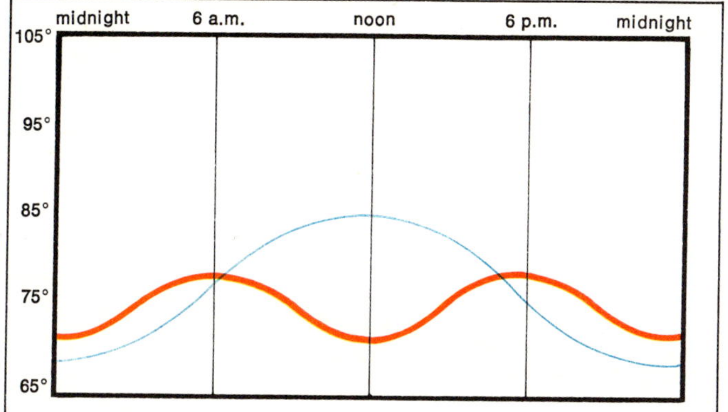

The body-temperature line (in red) of a very primitive amphibian, ichthyostega, that lacked internal temperature control, swings widely; the animal escaped cool of night and heat of day (blue) by taking to water.

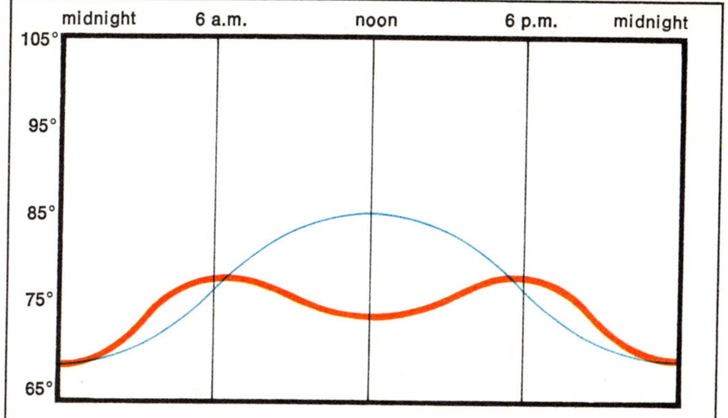

The body temperature of the early reptile romeria also changed with air temperature since it too lacked real temperature control. Romeria used sunshine to warm up in the mornings and shade at noon to cool off.

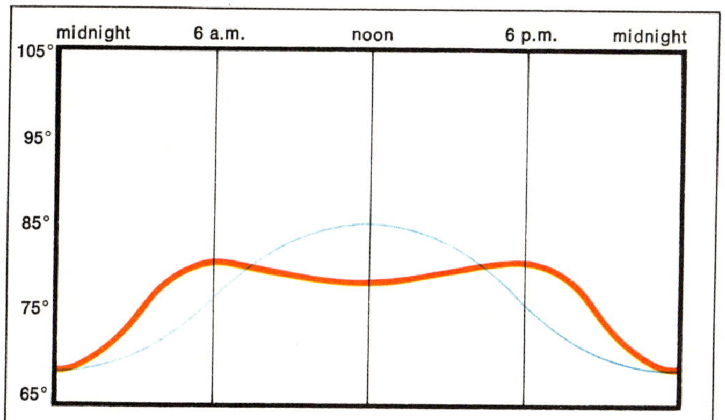

The beginnings of internal temperature control are charted here for the mammallike reptile thrinaxodon. Its temperature upon waking rose steeply; at noon its temperature did not need to drop so far for comfort.

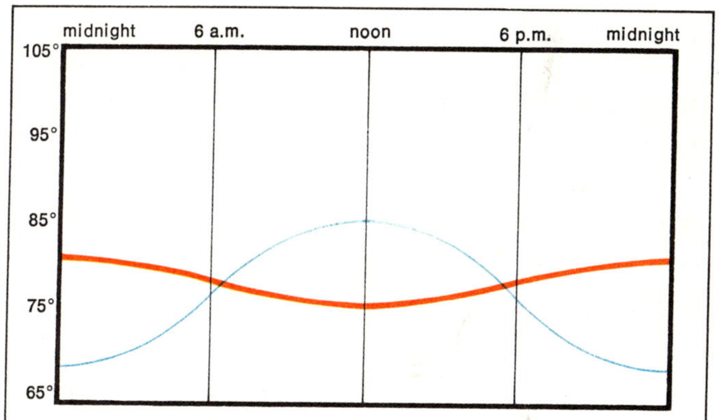

The primitive mammal pantothere had a body temperature that was not much influenced by outside temperatures. Its average temperature was 80.6°F., increasing the energy output needed for an active life.

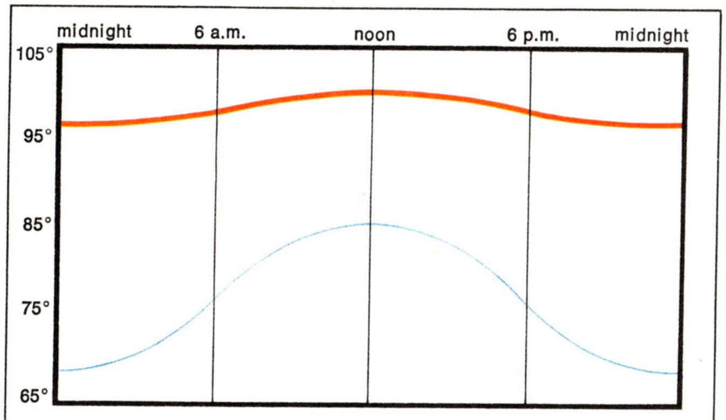

The temperature recorded for the primitive squirrellike monkey propliopithecus shows a major evolutionary improvement: body temperature, much higher than the air's, fluctuated markedly only in the heat of noon.

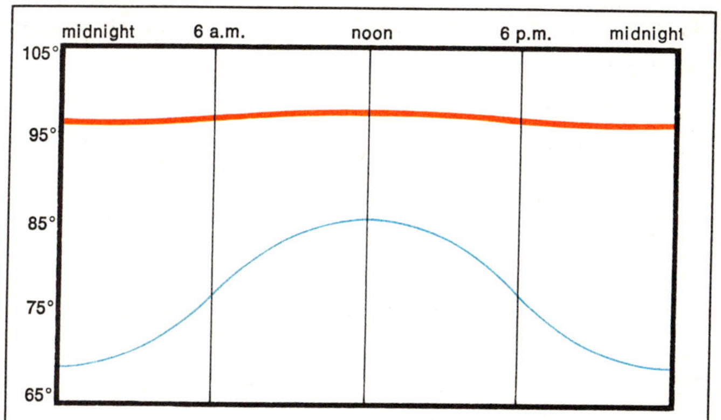

Man's chart shows an almost perfectly steady high temperature throughout day and night. Man, almost free of dependence on outside temperatures, is able to pursue an active life in almost any environment.

with the reptiles, the active ones called mammallike, whose general line of development suggests that they may have had a furry covering. Among the mammals, fur has developed into an excellent protection against cold. They can adjust the insulating power of their fur by making the hairs stand up to form a thick heat-retaining layer. Human "fur" is not much use for insulation any more, but the few hairs that remain still stand up dutifully as goose-pimples when the tiny muscles at their roots are told by the body that it needs more protection from the cold.

The second mechanism of temperature control, and apparently an ancient one, is shivering. It produces heat through muscular activity but does so automatically, without the conscious effort muscular activity normally involves. Shivering is common among mammals and has been reported in reptiles and insects. Some snakes apparently shiver to generate extra heat to keep their eggs warm. Pythons in New York's Bronx Zoo have been seen to coil around their eggs when the room temperature is low and then to contract their muscles in a spasmodic manner that resembles human shivering; the action keeps body temperatures from dropping as the air cools below 77°F. Even an insect such as a butterfly may shiver, contracting its wing muscles to warm them up before taking off for a flight on a cold day.

One temperature-control mechanism both warms and cools the human body: the blood-circulating system. Blood streams through the body, carrying heat from internal organs to the capillaries near the skin, where it can be dissipated into the cooler air. But if the body is already too cold, the blood flow near the skin is restricted so that less warmth is lost.

To protect against overcooling there is also a tricky arrangement in human arms and legs, similar to the device called a countercurrent heat exchanger in machines. These extremities lose heat rapidly—everyone's hands and feet get cold before the rest of the body. Because they are so thin their heat-dissipating surfaces are large compared to their heat-conserving bulk. To reduce this loss, the arteries carrying blood outward are deep inside the limbs, each paralleled by a nearby pair of veins. The returning blood in the veins picks up heat from the outgoing blood in the adjacent arteries, preventing all of this heat from being dissipated in the capillaries and carrying some back into the trunk of the body. The human countercurrent system is used, however, only when the body needs to conserve its heat. On the other hand, when cooling is required, the return flow of blood is switched from the countercurrent-exchange veins to another set of veins near the skin, where the returning blood can pick up no arterial heat. This changeover is actually visible—in warm weather the veins under the surface of the arms are noticeably larger than they are in cold weather.

How the countercurrent system of temperature control evolved is not known. It arose—apparently independently—in a number of animals, including man, his distant mammalian relatives the whales, and birds like geese that spend much time standing in cold water. And at least one fish, the tuna, has such an arrangement to reduce heat loss to the water passing through its gills and thus maintains a body temperature higher than its surroundings; as a result, the tuna is much more energetic than other fish and can maintain long bursts of high-speed swimming.

While all mammals use blood flow both to warm and to cool their bodies, they also have specialized schemes solely for cooling. Man sweats. The water released through pores in the skin evaporates and in doing so gets rid of heat. A few other mammals, such as horses, sweat, but many—dogs, for instance —achieve the same end by panting. How dogs' panting helps cool their bodies has only recently been learned. They rapidly pull air into their lungs through their noses, where the air is cooled by a secretion of water; the cool air then takes heat away from the inner surfaces of throat and lungs. Still other mammals have evolved a different technique for evaporative cooling: they moisten their fur by licking it.

The main center for controlling these temperature-regulating mechanisms is the hypothalamus, which is located at the base of the brain and functions somewhat like the thermostat governing the furnace in a house. It is extremely precise. When the temperature begins to drop or rise, the hypothalamus first orders a decrease or increase in blood flow. If the air temperature around an unclothed man decreases below 80.6°F., blood flow cannot compensate for heat loss, and the hypothalamus signals for shivering to provide internal warmth; at 87.8°F., blood cooling is insufficient for an unclothed man, and at that temperature sweating begins.

Uniform temperature seems associated with intelligence. The connection is too complicated to spell out here, but it is evident that the only animals with notably elaborate brains are the mammals and the birds, both of which have warm bodies. Moreover, only warm-bodied animals have the complicated behavior patterns that have meant so much to their survival. For instance, they take care of their young much more effectively than the cold-blooded and less intelligent reptiles. Only a few reptiles stand guard over their eggs, hardly any feed the hatchlings as most birds do, and no reptiles nurse their young or give them the long protective attention that man and most other mammals give their offspring.

It is safe to say that without warm bodies to encourage vigorous activity and the growth of intelligence some of the first true mammals, small, shrewlike creatures that appeared before the dinosaurs' reign ended, could not have given rise to the primate line that culminated in man. This active life found its first great challenge in the trees. Arboreal living is not for the stupid or the fumbling. To run along yielding branches and leap from tree to tree called not only for good vision and good balance but for quick mental computation. It also required hands and feet that could clutch branches securely.

Man's amazingly dexterous hands, his acute stereoscopic vision and his superior brain are clearly the legacy of agile tree-dwelling ancestors. A foreshadowing of the human hand probably first occurred in a creature resembling the lemurs, primitive primates that still inhabit the tropical forests of Madagascar. Modern lemurs live in trees, like squirrels, but instead of climbing by means of sharp claws as squirrels do, they use fingers and toes to grasp twigs and branches. Their thumbs and big toes are to a certain extent opposable to the rest of the hand or foot, giving a better grip and permitting the lemurs to pick up and manipulate objects. Modern monkeys, descendants of primates considerably more advanced than the lemurs, have hands approaching man's in their range of movement and general dexterity.

Arboreal life, developed only by the primates, has been a prime factor in the development of man's three-dimensional vision. The eyes of most mammals

Refining the Senses

Evolving from a line of vertebrate ancestors, man emerges with a highly refined sensory apparatus, the most complex and most important part of which is the intellect.

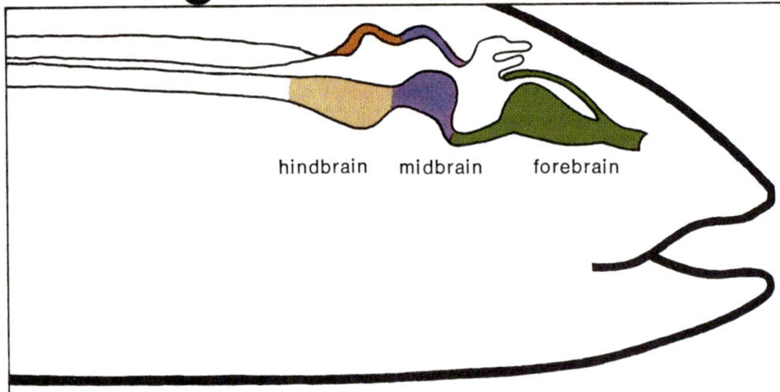

Man's brain has been built upon the fundamental structure shown in the fish: the forebrain (green), concerned with the sense of smell, the midbrain (purple) with vision, and the hindbrain (brown) with balance and hearing. Since smelling is so important to the fish, the forebrain is relatively large.

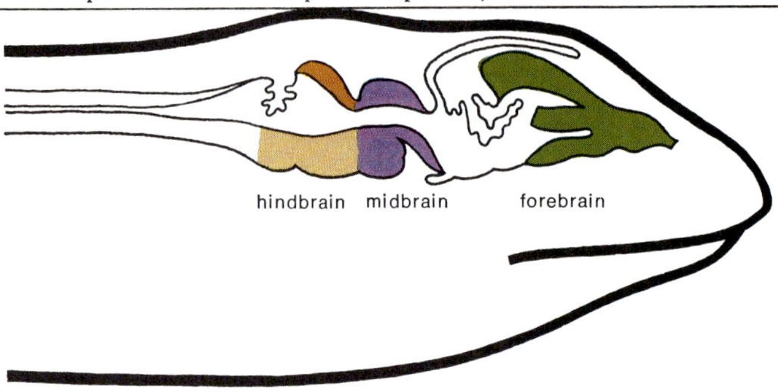

In the reptile the organs of hearing and vision have become more important, so both the midbrain and hindbrain are enlarged. Furthermore, the brain as a whole is becoming more complex. For example, the midbrain has expanded its role in coordinating the reptile's increased sensory activities.

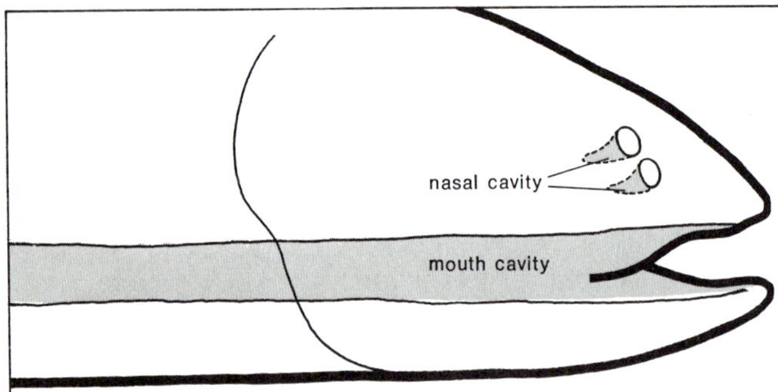

In the evolution of vertebrates, smelling is linked to breathing and eating. The fish's smelling and breathing organs are separate: in the species above, the sense of smell lies in four membrane-lined pockets (two of which show in this side view). These deal only with smell; most fish breathe by gills.

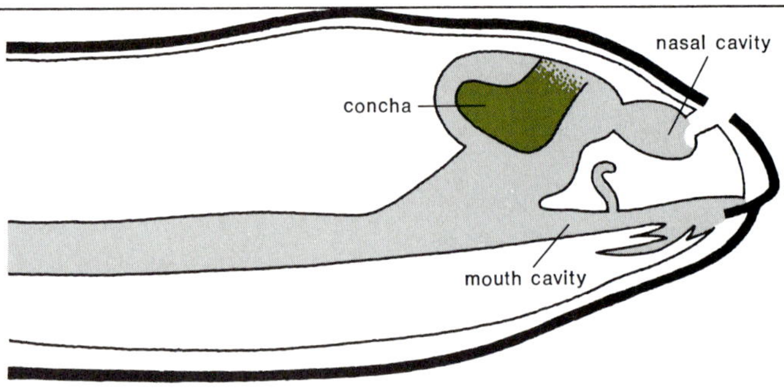

The reptile smells and breathes through one passageway (gray) opening into the mouth. Thus the reptile cannot chew and breathe simultaneously. But there have been advances: a membrane-lined concha (green) humidifies incoming air; the hook-shaped pocket above the mouth is a tasting area.

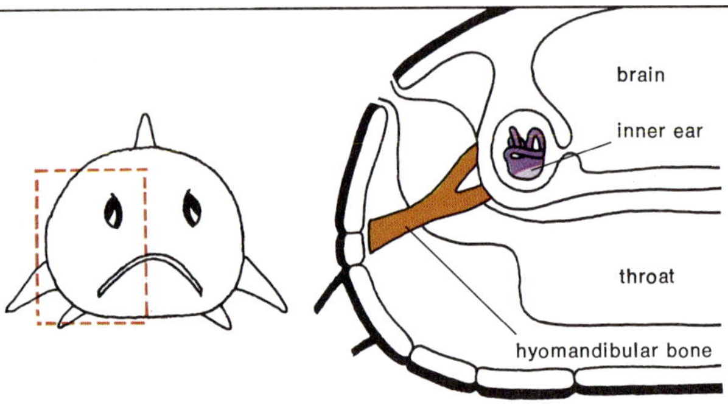

The early vertebrate ear served primarily as an organ of balance. As the cross section at right shows, in fish the organ consisted only of an inner ear (purple); a hyomandibular bone (brown) transmitted vibrations to the inner ear from the water, constituting the start of a hearing mechanism.

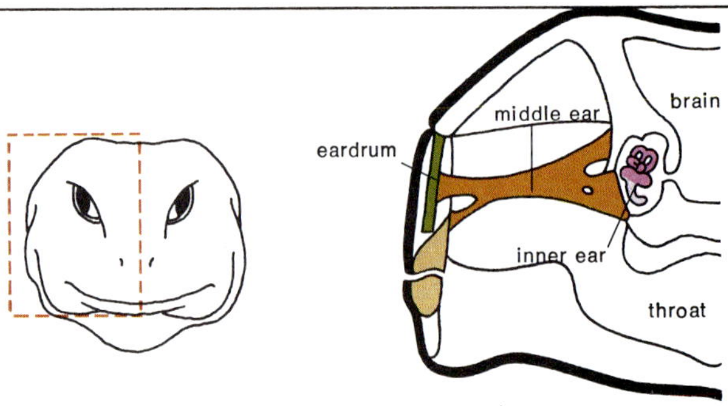

In the land-dwelling reptile the mechanism for hearing has become more intricate. Now an eardrum (green) transmits sound waves via a middle ear (which has evolved from the fish's hyomandibular bone) to the inner ear. Two bones (light brown) below the eardrum form the joint of the jaw.

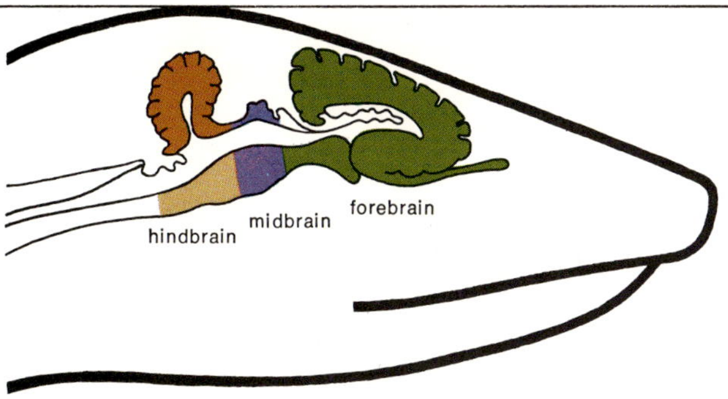

The mammal's brain is more complex, and sense coordination has moved from midbrain to forebrain. The forebrain has now developed a folded cerebrum on top, involving memory and learning. The hindbrain has also developed a cerebellum to coordinate increasingly complex movements.

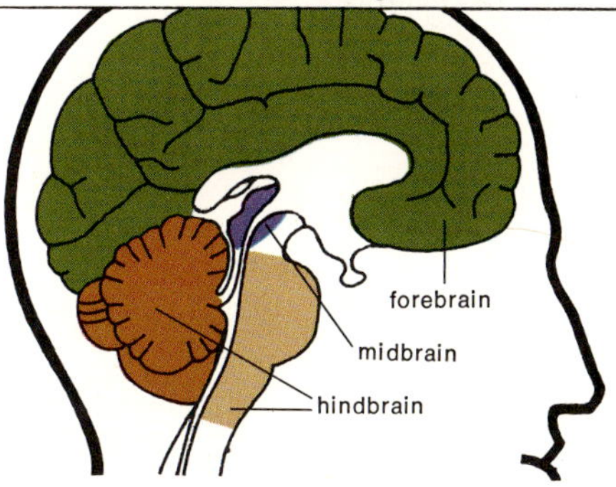

In man the midbrain and the hindbrain with its cerebellum have not increased greatly in size compared with the forebrain, which, with its cerebrum, now dominates the brain. Nearly all brain functions that are man's alone—most important, abstract thought—center in this part of his brain.

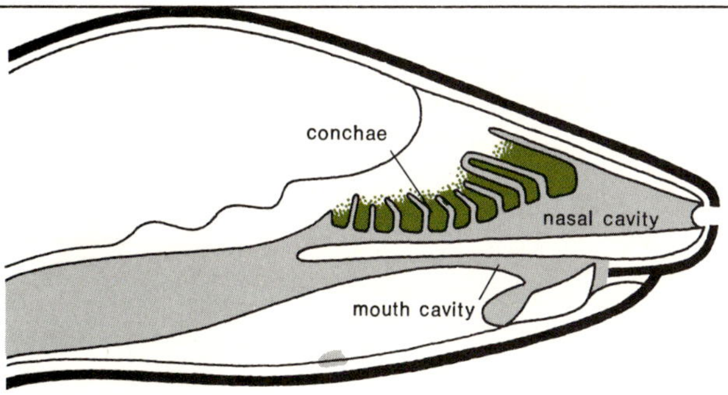

The mammal uses smell to identify prey, hence the large nasal cavity contains elaborate olfactory membranes. Numerous conchae (green) warm and humidify the air and help discharge excess body heat. A secondary palate between nasal and mouth cavities permits simultaneous breathing and chewing.

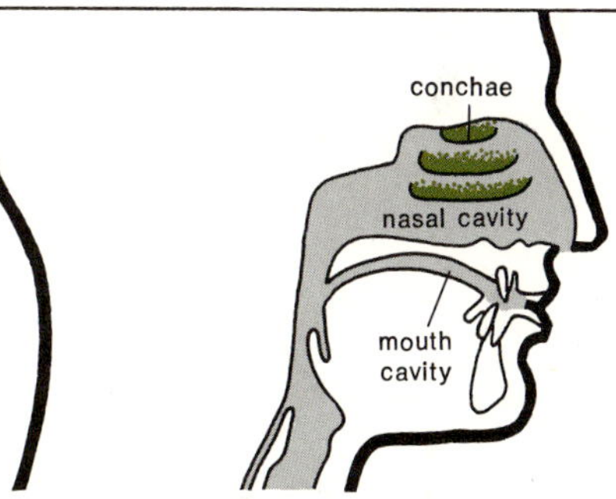

Man can chew and breathe at the same time. But the acuteness of his sense of smell has been reduced. The reason, many scientists believe, is that his tree-dwelling primate ancestors concentrated chiefly on vision rather than on smell to survive. Similarly, only three conchae (green) remain in man.

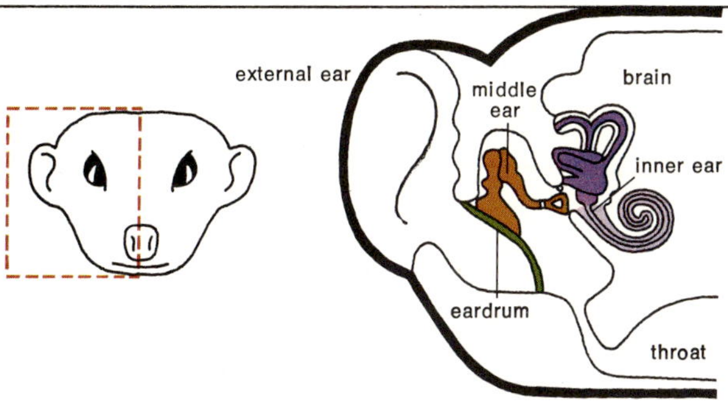

In the mammal the bones that formed the reptile's jaw joint have moved inward (brown) to become a part of a middle ear, which acts as an amplifier. The inner ear (purple) now has a spiral tube, the coiled cochlea (light purple). It is lined with a membrane whose parts vibrate to different frequencies.

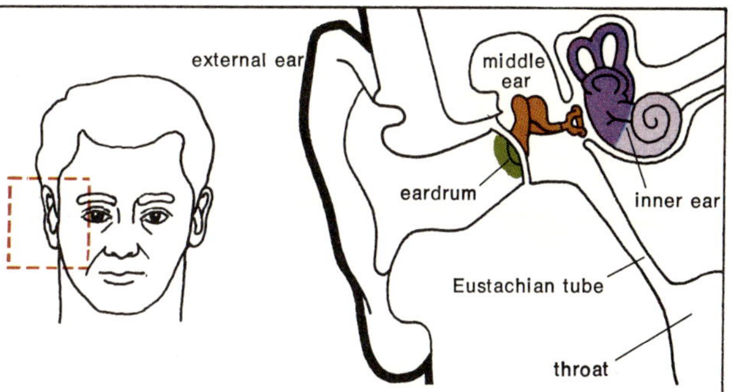

The ear of man is simply a well-developed mammalian ear, with an external ear (right) that collects sounds; an ear canal that funnels them to the eardrum (green); a three-part middle ear (brown); an inner ear (purple) that differentiates frequencies and then passes on the information to the brain.

are located toward the sides of the head so that the animal sees two separate pictures, both of them lacking in depth. An improved and more humanlike system of seeing can be traced to a primitive primate that may have resembled the tarsier, an odd little hopping animal, now living in the East Indies, that uses slender, fingerlike toes to cling to upright branches and stares at the world with large night-prowling eyes. Instead of being placed on the sides of the head these eyes have moved to the front of the face, where they point well forward, as do those of man and other higher primates. Thus their fields of view coincide, permitting the tarsier to observe its world in depth through stereoscopic vision. The eyes of monkeys and all other higher primates are also normally equipped with the fovea, the section in the eye's retina that gives a tiny area of sharp, colorful vision in the center of a much broader but fuzzier and duller view furnished by the rest of the retina.

The great improvement in vision that came with life in the trees also stimulated the growth of the brain. The brain, in fact, has apparently developed over billions of years largely in response to the demands of the senses, for the signal of a sensation must lead to a reaction inside some control center. Eyes pick up visual images but only the brain sees the image. The earliest sense to arise was touch; even single-celled organisms used it to gather their food, "swallowing" particles they brushed against, and it might be argued that the very appearance of the first life on earth depended on the chemical touch between nonliving molecules in the primordial waters. Touch, highly refined, remains an important sense to man —it enables the blind to read Braille, distinguishes wool tweed from silk satin, and unfortunately helps burglars crack safes (they sandpaper their fingertips to bring the touch-sensitive nerves closer to the surface of the skin). But touch works only on contact —at zero distance between the sensor and the thing to be sensed. As a tool for finding food and recognizing friends or enemies, touch is far surpassed by smell, which works at a distance.

Smell is a specialized form of touch, since it depends on contact between odor molecules in air or water and sensitive nerve endings in nose and mouth. By the time fish had evolved, the smell sense was well developed and the nerve cells that received smell signals had developed into olfactory bulbs at the forward end of the little brain. Smell is extremely refined in such modern fish as salmon, which depend on it to find their way hundreds of miles up a river to reach the spot where they will lay their eggs. Man's sense of smell is also more delicate than is generally recognized—he can detect as little as 32 millionths of a billionth of an ounce of the substance musk. (However, man's sensitivity is far surpassed by that of the male gypsy moth, which can smell a female seven miles away, responding to four millionths of a billionth of an ounce of sexual attractant.)

From the fish, man also gets his sense of hearing. It apparently developed in the early jawless fish as a balancing device, a hollowed curve within the skull filled with fluid and with cells that responded to movements in the fluid. All this device did was help keep these fish on an even keel; it was not useful for hearing. Later fish, however, evolved an air sac that was used to regulate buoyancy, and this sac converted the original balancing organ into a hearing device. Pressure waves—sounds—striking the sac disturbed body fluids inside the fish, and movement of these body fluids affected the balancing organ.

This hearing arrangement worked well for fish, but when amphibians had to hear in air they ran into the problem that engineers call an impedance mismatch. Sound waves in air now had to cause a response in a listening device that was filled with fluid—as it still is in man. The human nerve endings that transmit sound to the brain are submerged in fluid in the cochlea of the inner ear. The solution to this air-fluid problem was the eventual development of three of the most delicate bones of the human body—the hammer, anvil and stirrup bones of the middle ear—which, attached to a new structure, the eardrum, convert sound from the air-pressure waves vibrating the eardrum to fluid pressure waves in the cochlea. The human middle ear is the end product of long and complex evolutionary steps: its cavity evolved from one of the gill slits of fish, and its three crucial bones evolved from bones of the fish's jaw.

The twists of evolution that converted a fish's balancing organ into a human hearing device gave man an extremely delicate sense. And hearing eventually became one of the most crucial of human senses—a child born totally deaf has great difficulty in learning to speak, and his lack of ability to communicate may cut him off from human society. However, not hearing or smell but vision led to the great expansion of the brain that elevated man over all other animals.

For most mammals smell is the all-important sense. They snuffle their way through life on the ground, depending on odor to lead them to food and to warn them of danger. But as the primates evolved, their sense of smell declined as their vision improved, and their brains were modified to encompass the flood of information passed to them by the eyes. The combination of good hands and good vision, constantly employed, helped give the monkeys comparatively large brains, much in the cerebrum, the frontal portion where intelligence resides. The skull bulged upward to accommodate the brain, giving the monkey face that human look that is fascinating to some people, disquieting to others.

Primitive primates resembling lemurs and wide-eyed tarsiers made enormously important contributions to the development of man's body, and his closer cousins, the apes, carried the progression further. In structure and ability the brains of modern apes definitely approach those of man. They are built like human brains with a generous amount of "gray matter" on the wrinkled surface of the cerebrum. With this humanlike brain comes the start of a humanlike memory and reasoning power. Captive chimpanzees are particularly adept at solving problems devised for them by animal psychologists.

With all their humanlike characteristics, however, the apes never evolved the one crucial trait that makes the human body unique on earth—the ability to walk upright through life on two feet. But some ancient relative of theirs did.

Erect posture put the man-ape on the road to becoming a true man. It set his hands free to develop greater manipulative ability, placing still greater demands on vision. And the interaction of all these developments stimulated further growth of the brain. By and large no further major skeletal improvements were needed, only minor changes of proportion such as lengthening of the legs. The physical construction of man's body—which began over a billion years ago with a simple, soft-bodied creature with a tube extending through it—was now complete.

The Expressive Face

These drawings show how facial expressions—and the muscles that produce them—evolved from a fish's blank mask to the eloquence of man's ever-changing face.

The dead-pan coelacanth, modern descendant of a primitive fish, lacks the facial muscles that enable complex creatures (below and opposite) to express fear and hostility and, among higher animals, even joyousness.

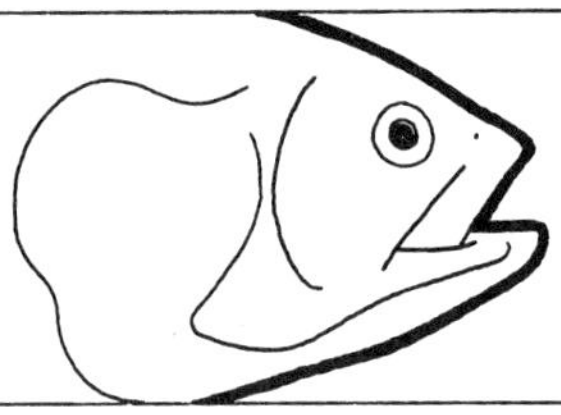

One reptile capable of facial communication, the frilled lizard, uses muscles (fine lines) modified from those of the typical lizard. By spreading its folded ruff (far right) when endangered, it has two expressions:

1. Normal
2. Threat

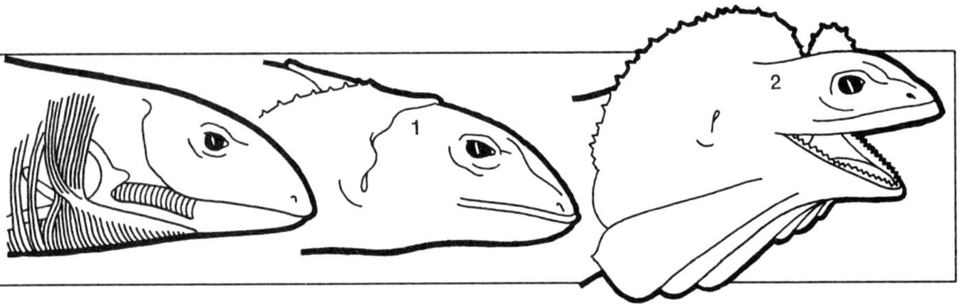

As mammals evolved, muscular tissues crept forward to cover the face with the mobile mask characteristic of the group. The opossum shown here, a relatively primitive mammal, can use its simple muscles to communicate a threat when startled, to express fierceness when alarmed, and to "play possum" like an unpalatable corpse:

1. Normal
2. Strong threat
3. Mild threat
4. Feigned death

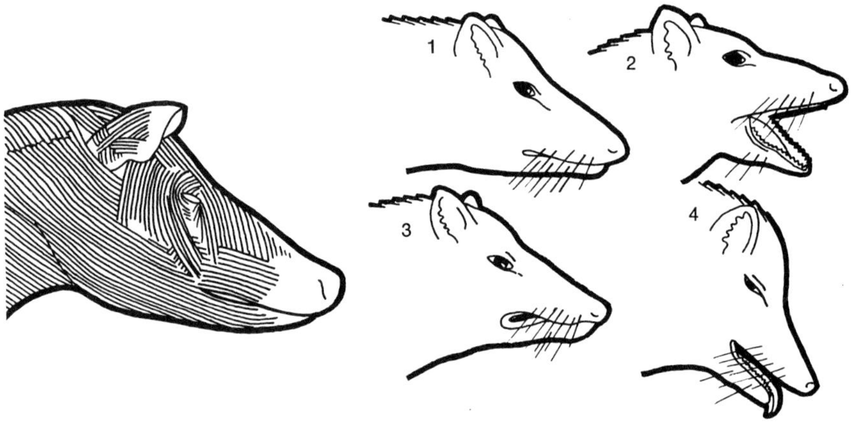

In the primates, facial muscles become increasingly specialized, permitting more complex combinations of cheek and mouth positions to impart the precise messages a social life requires. A surprised rhesus monkey first reacts with a mild threat like the opossum, but when it learns whether it has been startled by foe or friend, its face can convey many other meanings:

1. Mild threat
2. Despair
3. Anger
4. Lip-smacking (cautious greeting)
5. Grimace
6. Normal
7. Horror
8. Strong threat

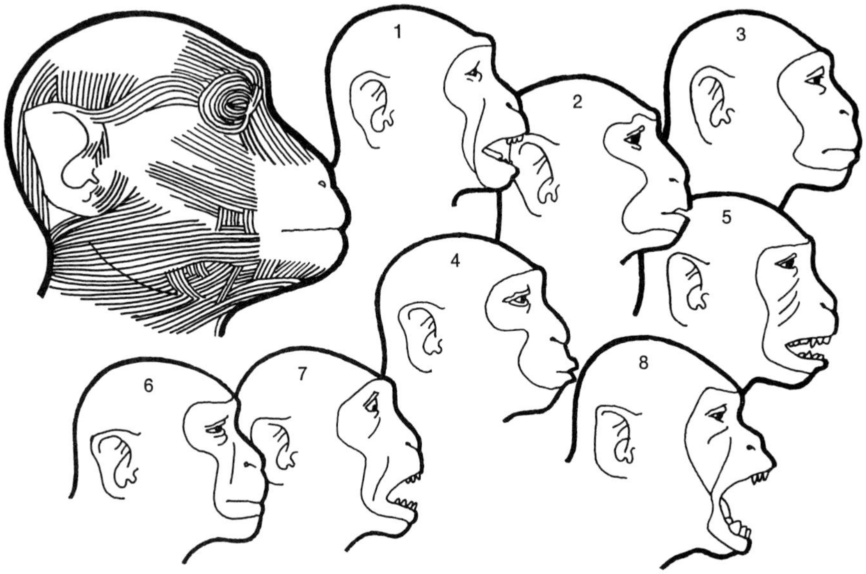

The very intelligent chimpanzee communicates not only from necessity, but often simply because it enjoys exchanges with its fellows. It is equipped with a far richer complex of facial muscles—especially around the eyes, brow and mouth—than most other primates. A chimp can convey not only such basic reactions as anger or terror, but a variety of emotional and factual information required by a life of a highly sociable character:

1. Wailing sadness
2. Fright
3. Astonishment
4. Attention
5. Frenzy
6. Hooting excitement
7. Enjoyment
8. Normal
9. Hilarity
10. Anger
11. Terror
12. Grinning amusement

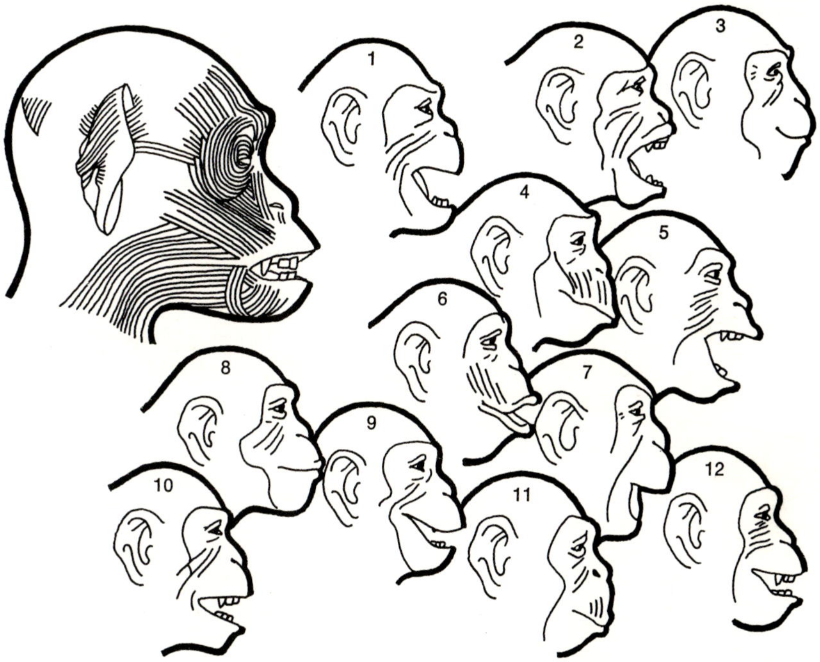

Many specialized muscles—almost all set in pairs—control man's eloquent repertoire of facial expressions. Two pairs run down the sides of the nose to raise lips and nostrils in disbelief (drawing 13), and another pair raises the mouth in a smile (3). The band that knits the brows is one of the few unpaired expression-producing muscles.

1. Silent pain
2. Skepticism
3. Hilarity
4. Flirtatious interest
5. Mocking inquiry
6. Normal
7. Joking threat
8. Amusement
9. Surprise
10. Sharp pain
11. Playfulness
12. Eagerness
13. Disbelief
14. Concentration
15. Fear
16. Bored cynicism
17. Rage

Chapter Five: The Power of the Group

Bold, strong individuals, Africa's oryx antelopes nevertheless live in sociable groups. Here a herd streams across the grassland.

Every man likes to think of himself as a unique and independent individual, separate from all other humans and able, if circumstances demand, to get along without them. He cannot. Humans need to be with other humans, that is, to be members of a human society. Only under very exceptional circumstances can a man live more than a few weeks without the vital benefits—food, shelter, protection, cooperation, information and simple companionship—that society alone is able to provide.

The most famous man who lived in total isolation for a while was Alexander Selkirk, the sailor model for Daniel Defoe's fictional character Robinson Crusoe. But Selkirk was able to survive his lonely experience on a deserted isle because it took place in an almost ideal spot. Juan Fernández island, off the coast of Chile, where he was marooned in 1704 after a quarrel with the captain of his ship, was reasonably fertile, with a temperate climate and no dangerous beasts. It was stocked with goats, which he managed to catch for food and clothing. And he had with him his personal possessions, including such helpful artifacts of an advanced society as a musket and ammunition. Even so, he suffered horribly from loneliness (the castaway's man, Friday, was Defoe's invention), and when he was rescued after four years and four months, he could speak only incoherently. In a less favorable place, without his gun, clothes and other inanimate support from human society, even the hardy and self-reliant Selkirk would have fared far worse.

But beyond a man's need for association with fellow individuals, there is the requirement for another kind of association, internal rather than external. Each human is made up of a society of trillions of closely cooperating cells, some of which look and behave like the independent one-celled animals that their ancestors once were. And even these cells, which make up the human body, may each also be a kind of society, the product of lower levels of association among still more primitive single-cell bacterialike units of life.

So man, who believes himself individual, is both an obligatory participant in a higher society and the product of earlier levels of association. His dependence on social organization costs him individual freedom but it pays him back many times over in the great power that comes with group life. Alone, a man may be physically and mentally superior to any other animal, but his individual advantages do not make him dominant; only with the development of human society did man come to rule the earth.

Man's social organizations—family, hunting band, village, tribe, nation—arose from his own special qualities, and they cannot be said to have evolved from the group living engaged in by other animals. Yet many of the elements of human society have now been shown to exist in other societies, so that animal and even plant life explain in many enlightening ways characteristics of man's life. Thus it has become possible to explore the development of social organizations on earth over millions of years, partly by studying the clues to organized life discovered in fossils and partly by deducing the early patterns of association from those that are still in existence today among both primitive and advanced organisms.

In a sense, life depends on organized association. The first living things appeared when certain chemical substances were organized into a pattern that enabled them to reproduce themselves, generation after generation *(Chapter 2)*. But for some two billion years, each such microscopic grouping of chemicals was an independent unit of life, surviving without assistance from its neighbors. According to a 19th Century theory of the evolution of animal cells that has been restated in recent years, the first step toward society came when two dissimilar units of life joined forces and each became dependent for survival upon the other—a kind of existence scientists call symbiosis. This great leap toward complex, advanced ways of living occurred somewhat less than one billion years ago, when the waters of the earth were populated by very simple single-celled bacterialike microorganisms. There were many kinds, for they had been slowly evolving for more than two billion years. Some of them swam by means of whiplike tails. Others floated passively in the water, absorbing dissolved food through their body walls. One kind, the blue-green algae, had gradually acquired the ability to make food out of water, carbon dioxide and the energy of sunlight through a process called photosynthesis. The free oxygen released during this process was slowly accumulating in the atmosphere where none had been before.

This buildup of oxygen brought about a major crisis in the history of life, but a crisis that made possible the evolutionary process leading eventually to modern man and his society. To most of the living creatures of that remote age, free oxygen was a deadly poison. Some species doubtless disappeared when oxygen invaded the water they lived in. Others retreated into oxygen-free mud, where their descendants live today. A few managed to adapt to the new, dangerous gas. They not only prevented it from damaging them but also utilized the energy released when it reacted with their food—carbon-containing sugar compounds present in the water. Since this carbon-oxygen reaction releases much more energy than do earlier life processes, these users of free oxygen became the most efficient things on earth.

Presently an extraordinary thing happened. According to the symbiosis theory of cell evolution, a large bacterium accustomed to the old-fashioned way of living, without oxygen, was joined by one or more of the new-style oxygen-users. These oxygen-users entered the body of the host bacterium but did not harm it, nor did the host digest its guests. From then on, the organisms lived together in the even-handed partnership of symbiosis. The large cell engulfed or absorbed carbonaceous food, only partially utilizing it as before, while the smaller guest cell (or cells) inside the large one combined the partially broken-down food of its host with oxygen to produce additional energy for both.

This rich supply of energy from free oxygen made the first such symbiotic partnership more efficient than other single cells without partnerships, but the new life form's ability to move around was limited (it may have merely floated). That fault may have been corrected when many threadlike bacteria—similar to modern spirochetes, whose entire bodies wriggle quickly—attached themselves to the outside of the composite cell. They derived energy and nourishment from the host while the vigorous motion of their bodies enabled what had now become a triple partnership to move rapidly through the water in search of food. The new mobility, which was backed by plenty of energy, made such composite cells the terrors of the one-celled world.

Whipping tails, or flagella, which still act like outboard motors for many microorganisms, may not be the only contributions that threadlike bacteria made to the composite cells. While the flagellum stayed at the periphery of the host cell, where it acted as a propellant, part of it may have moved deep into the host and contributed to the evolution of the nucleus, which came to control the cell's reproduction.

Most of these ideas are being hotly debated. The majority of scientists maintain there simply is not yet enough evidence to be sure of this explanation of higher cell evolution. But however the nucleus developed, it is so important that biologists make a distinction between prokaryotic (prenuclear) bacterialike cells, which evolved during the long ages while the earth's atmosphere slowly acquired its free oxygen, and the eukaryotic (truly nucleated) cells, which were made possible, if the symbiosis theory is correct, when prokaryotic cells of three kinds joined in partnership. In any case, the nucleated eukaryotic cells became dominant, and all modern animals, including man, are descended from them.

Although this eukaryotic experiment at group living led at first only to single-celled organisms, it proved very successful. The eukaryotic cells were quick to specialize, evolving into innumerable forms to exploit all available ways of getting a living. Some swam fast, some more slowly; some crawled on solid surfaces, others sat still and waited for food to come their way. Most remained microscopically small, but a few grew to a size big enough to have been seen with the naked human eye, if there had been one. Some of these nucleated single-celled forms became extremely complicated. Modern examples have senses of taste, touch and sight (or at least light-sensitivity). Some kinds, such as paramecia, have well-planned mouths and a digesting and eliminating system. They swim by means of innumerable cilia (small flagella) all beating in unison. When an obstacle is encountered, the cilia go into reverse, making the organism back away. So there must be something like a nervous system to synchronize their cilia.

One obstacle these fierce little predators did not overcome was their small size, an inherent limitation of one-celled organisms. They get their oxygen supply from the water by simple diffusion through the cell membrane. The larger the cell becomes, the more oxygen it needs and the more trouble it has getting the oxygen to its interior. Conceivably, some system could have evolved that would carry oxygen into the cell so efficiently that much larger one-celled animals would have received all they required; but so far as is known, such an adaptation never appeared. Instead, to gain the many competitive advantages of larger size, some one-celled animals used another stage of association. This time it was cells of the same species that banded together, became interdependent and formed multicelled animals. The superorganisms that came about in this way are called *Metazoa*. Ants, elephants, mice and men are metazoans. They are all superorganisms of associated cells.

Most biologists have little to say about the origin of the metazoans. They do not know positively how they arose and do not expect ever to have anything better than fairly convincing hypotheses. The great event occurred deep in the Precambrian era, which ended about 600 million years ago, and all the likely participants were microscopically small and so soft-bodied that there is little chance they left meaningful

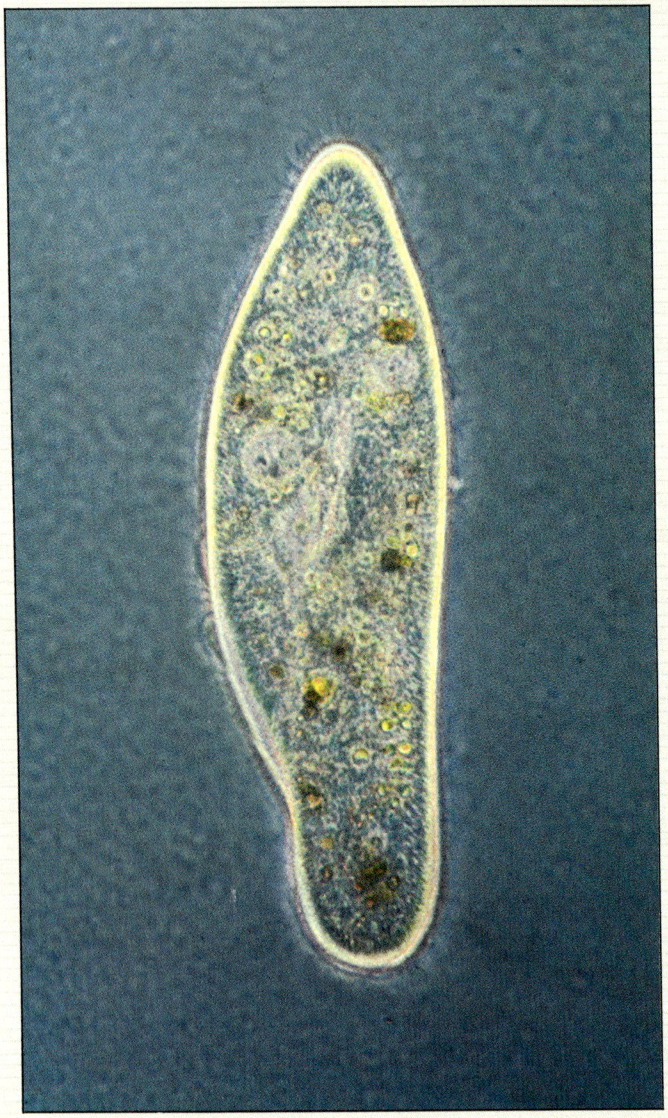

The structure of this single-celled microorganism—a modern paramecium believed similar to ancient forms—exhibits specialized features as do cells of the complex human body. The paramecium has two nuclei to control reproduction (a large one right of center, and a small one, not seen, next to it) and countless hairlike cilia, some to propel it through water and others to sweep food into its body.

fossils. Certainly none have been found. The best the biologists can do is to study the simplest modern metazoans in hope of determining how the first of them arose from one-celled animals.

One idea holds that the first metazoan developed out of large single-celled forms that were covered with small whipping cilia and had several or many nuclei. Since one nucleus is all that is needed for a cell, it is possible that some of these ciliates acquired partitions to separate their nuclei. A cell partitioned into sections, each with its own nucleus, is in effect a multicelled organism.

A more widely held hypothesis maintains that the metazoans originated not by the division of one cell into many but by the association of one-celled flagellated organisms into colonies. Such colonies of one-celled organisms exist today. In some cases the cells remain essentially unchanged; if separated, they may lead independent lives and form new colonies by dividing normally. In others the cells have surrendered their independence and assumed a function in the multicelled organism that makes normal life impossible outside it.

The best-known colonial form that appears to have some specialization among its cells is volvox, a beautiful, green, slightly elongated hollow sphere about three hundredths of an inch in diameter that swims through the water while spinning merrily. Volvox contains chlorophyll and can therefore be classed as a green plant, but at its level of life there is no firm dividing line between plants and animals. Many lowly organisms act like plants in depending partly on photosynthesis but otherwise move and feed like animals. So in discussing volvox it is quite proper to ignore for the time its plantlike characteristics.

Volvox is made up of a single layer of cells that are almost exactly like free-living flagellates, single-celled organisms that swim by means of flagella. Normally the cells are firmly fixed in the volvox sphere, with their flagella pointing outward, but if one of them is detached it swims around happily as if it were wholly content to live alone. However, it cannot reproduce, and after a while it dies. Life as a volvox cell has obviously cost it something of the independence of a single-celled organism.

The volvox organism, in fact, controls its constituent cells in several ways. It makes their flagella beat in unison so that they move the sphere through the water and periodically reverse its spin. Only certain cells take part in reproduction. Volvox demonstrates to a degree the two main characteristics of any metazoan. Its cells have specialized, even if only slightly, and they all cooperate for the welfare of the organism as a whole—in the same way that the cells of the human body do.

Not all metazoans are necessarily descended from anything like volvox. The habit of forming colonies is not uncommon in the one-celled world, so some scientists believe that the border between independent single cells and organized groups of cells was crossed more than once, making it possible for different kinds of multicelled animals to descend from different kinds of colony-forming cells. Indeed, some humble forms living today have not made up their minds about which side of the line they belong on. Among them are certain slime-mold amoebas, which spend part of their active lives as independent cells that look like other amoebas. They normally live in the soil and crawl slowly through it engulfing bacteria and reproducing by simple division. When all available bacteria have been eaten, the amoebas abandon their individual free life and behave in unison like the cells of a metazoan.

The slime-mold cells will act out their crossing of the evolutionary border on a laboratory culture plate where they can be watched. By the tens of thousands they can be seen to stream toward central points, making blobs that are easily visible to the naked eye. From each, a peak rises up, falls sideways and forms a sluglike creature as much as three fiftieths of an inch long that crawls slowly toward light and warmth. If it were in its native soil, it would normally seek the surface.

After crawling a while, the slug upends itself. Some of its cells form a base firmly attached to the surface. Others make out of their bodies a slender hollow stalk. The remaining cells flow up the stalk, turn into thick-walled spores, gather in a spherical mass and wait for better days. Some of the amoebas have been sacrificed; their dead bodies have gone to build the base and stalk. But the others get a chance to be widely distributed. If they had stayed in the depleted soil as individuals they all might have died of starvation. However, once the spores reach a proper environment, such as fresh, moist soil, they break open. Out of each flows a free-living cell hungry for bacteria.

The earliest metazoans were not at all like man, or any of the other higher animals or plants, but they had enormous potential for improvement. The future of life was theirs. Released from the one-celled body plan, they could form large structures of many advantageous shapes. Their constituent cells could specialize to perform particular duties, such as forming a protective outer skin. Large numbers of them could act in unison to change the shape of the organism or to move its limbs or tentacles.

As the metazoans became more complex, most of their cells lost the versatility of their independent ancestors and concentrated on a specialty like the members of any complex society. Muscle cells developed the ability to lengthen and contract, transforming latent chemical into active mechanical energy. Gland cells devoted themselves to producing a single secretion. Some cells stored reserves of fat; others became connective tissue that held the organism together.

Before the metazoans could grow big and become fast-moving, they needed better communication among their parts. When nerve cells evolved and specialized in carrying electrochemical messages, some of them gathered as a sort of computer, the brain, where information from the senses was received and analyzed, and from which commands were sent to outlying parts of the organism. The culmination of this development is, of course, man, whose highly organized brain is the most distinctive part of his body and the source of his dominant position on earth.

Body cells do not vary much in size, so the larger a metazoan is, the more cells its body is apt to contain. A large elephant may have something like six quadrillion (6,000,000,000,000,000) cells. Man, although he can be classed as the most complicated of the metazoans, gets along very well with about 60 trillion (60,000,000,000,000).

Among these human cells are some reminiscent of man's ancient origins, stemming from the time, perhaps a billion years ago, when all cells lived independent lives. Lining the air passages that lead to human lungs are cells whose whipping cilia move dust and other foreign particles toward the mouth and keep the passages clear. Their cilia are not different in any significant manner from the whipping tails of the one-celled ciliates.

Most of the body's trillions of cells normally live out their lives in complete subordination to the larger interests of the body as a whole, but occasionally one of them reverts to an ancestral urge to reproduce independent of body control. Since the body supplies everything the cell needs for growth, the maverick multiplies without limit and eventually kills the body by clogging vital organs with masses of useless cells. This is cancer, a cellular rebellion reminding us that our bodies are social organizations held together by laws that can be flouted only at the risk of death.

All animals of any size are metazoans—societies of cells—but the process of grouping together for more effective action did not stop with them. The next step was the formation of societies composed of many individual metazoans. The first creatures to make this evolutionary leap on a grand scale were the insects.

Appearing on earth at about the same time as the reptiles, over 300 million years ago, the insects quickly spread over the land and evolved into innumerable forms, both plant-eating and meat-eating. Their mastery of flight gave them a great advantage, but their heavy external skeletons contributed toward making growth difficult and keeping them small. Perhaps they could have overcome this size limitation by physical modification of their bodies, but they did not. Instead, some of them did something analogous to what certain protozoans had done almost a billion years before, when they evolved into the metazoans.

The individual insects did not stick together to form larger bodies, like the cells of the metazoans. Insects are much too complex for that. The individuals remained physically separate, but they acquired the ability to act in closely cooperating social groups containing many members. Insect societies can be likened to fairly large animals. A large colony of one ant species, for instance, may have as many as 22 million members that together weigh more than 40 pounds, but the colony acts like a single superorganism and can do many things impossible for individual insects. These strange and wonderful societies preceded by perhaps 50 to 100 million years the time when another metazoan, man, would form his own superorganisms, human societies, and come to dominate the earth.

No one who watches an anthill can fail to admire its extraordinary organization. Streams of ants issue out of the nest at an orderly pace, often along roads that they have cleared; the ants return with prey or other food; the nest itself is carefully built and administered, with guards at its door and the area around its entrance neatly policed. These characteristics suggest orderliness, discipline, planning, provision for the future, all regulated by an invisible force—an example for human behavior. A similar conclusion could be reached by watching almost any kind of ant, termite, social wasp or honeybee—all are insects and all show that eerie unity that makes the thousands of individuals in a colony behave with a self-subordination that seems to make them similar to the cells of a metazoan.

Of the four modern kinds of insect that have attained true social living, the oldest are probably the termites, which descend from cockroachlike ancestors and can fairly be called social cockroaches. The other three—wasps (and their relatives the hornets), ants and bees—are all rather closely related, descended from primitive wasplike creatures. The habits of the four differ greatly. Social wasps are winged predators that get some of their food by capturing other insects or spiders. Many ants are ground-living predators, but others are vegetarians; some have taken to farming of a sort, even to keeping insect cattle. The ethereal honeybees, which charm everyone who studies them, support their elegant colonies entirely on pollen and nectar extracted from flowers. Many termites specialize in eating wood, as all too many homeowners have discovered.

The customs of social insects vary as widely as their means of making a living, but most of the colonies are started by a female, or queen. She has wings that enable her to fly a considerable distance from her home colony and mate with a winged male member of another colony, thereby avoiding inbreeding. In some cases she mates with several males and stores enough sperm to fertilize her eggs during many subsequent years of egg laying.

A typical queen ant breaks off her wings soon after mating. She selects a patch of suitable soil, a rotting log, a cavity under a stone or a piece of bark, and burrows into it or under it to excavate a small chamber. She seals herself into it and waits almost motionless while some of her eggs mature in her abdomen. She extrudes them and when they hatch tends the tiny, soft larvae as devotedly as any human mother. She feeds them with secretions from her jaw glands—equivalent to milk—and sometimes with unhatched eggs. All this time she normally does not eat, but the large wing muscles in her thorax dissolve in her blood, helping her fat reserves to keep her alive and feed her young. If she cannot feed all the larvae, she cuts up a few as food for the others.

The first larvae turn into pupae, often wrapped in silken cocoons, and later emerge as minims: tiny, sterile female workers. In spite of their tiny size, they know exactly what to do. They issue out of the nest in search of food for themselves and the queen. If they are successful, the young colony prospers. The queen lays more eggs, and this generation grows into larger, full-sized workers. More tunnels and chambers are excavated. The queen, her mothering duties taken over by the large and small workers, turns into a passive egg-laying machine. She is fed every few minutes. Her abdomen becomes enormous; she may lay thousands of eggs in a single day, millions during her lifetime.

The mothering behavior of the queen is not very different from that of the females of other, nonsocial insects that take good care of their young. It is the behavior of the workers that is remarkable. There is no one to teach them what to do—the queen does not —but each species has its own pattern of behavior. The workers know how to build the nest, which may be very elaborate, with canny provisions for ventilation and insulation. They know in some way how to take care of the queen and the young, how to forage for food, how best to defend the nest against invaders. Most of these duties they perform cooperatively. They have no leader; the queen does nothing but lay eggs. At certain times of year winged males and fertile, winged females are produced to fly off to start new colonies.

How does the anthill govern itself so effectively? Or the wasp nest, the beehive or the termite colony? Instinct regulates most insect behavior. A worker confronted by larvae automatically provides food, operating according to a hereditary scheme that is built into the cells of its body. But instinct alone cannot explain the complex patterns of cooperation among insects. When a worker ant is attacked by a predator, soldier ants or other workers quickly arrive on the scene—even if they have been at some distance from the incident, and could not have seen it (many of them are blind in any case). How could they have known one of their colony was in danger? There is obviously some form of communication among insects, a system of signals that enables them to pass information from one to another.

The social insects have no sound-wave language capable of conveying detailed commands or data. Sounds, indeed, are not important to most of them. Vision, when present, cannot be used in the darkness of a nest. All social insects touch one another frequently, but for limited purposes such as begging for food. There is no known electrical communication between individual insects as there is between the cells of the metazoan. About the only remaining means for communication available to insects is the chemical sense of smell or taste.

This chemical sense is apparently the secret of insect society. Its existence had been known for many years, but only in the 1950s did extensive research demonstrate the extent to which social insects possess the ability to secrete and to respond in very sophisticated ways to a variety of chemical compounds, some of which are barely discernible to humans. These chemical signals are secreted by all colony members, and even by unrelated insects that have sneaked into the nest. These substances, called

From Independent Cells to Human Society

Single-Unit Life

First simple independent organisms of a type called monerans

bacteria
blue-green algae

Multiple-Unit Life

Groups of simple forms among a type called protistans

amoebas
parameciums
slime molds
most algae

Plants

mosses, liverworts
ferns
naked seed plants
flowering plants

Animals

All organisms made up of interdependent and highly complex specialized cells known as metazoans

Fungi

molds
mushrooms
yeast

Animal Societies

Complex groupings made up of individual metazoans

Insects	Mammals
termites	oryx antelopes
ants	prairie dogs
bees	musk oxen
wasps	monkeys
	apes
	man

As the chart above shows, social organization is not a uniquely human condition, nor was man the first to discover the advantages of banding together. From the earliest living things—simple, single-unit independent organisms—there arose some multiple-unit groupings. Some cells in these groups became interdependent and developed into multicelled green plants, fungi and the animals called metazoans. Nearly all metazoans engage in some social behavior, and a few lower forms like ants and termites live in rigidly organized groups. But only man has developed to a high degree schemes of cooperation that combine the flexibility of the individual acting alone with the power of the group acting together.

pheromones, act singly or in concert to trigger instinctive actions, thus controlling the behavior of the members of the colony; in some cases the pheromones also alter the functioning of insects' bodies or modify the bodies themselves. Often the chemicals are spread through the colony by the members' habit of feeding each other with the contents of their crops or digestive tracts.

A simple illustration of social insects' odor communication is the alarm signal used by most ants. When a foraging worker of one species, for example, meets a member of another colony or an insect predator, it looses minute amounts of volatile secretions from glands in its jaw and abdomen. When the scent reaches nearby workers they stop whatever they are doing and move toward their aroused fellow. If the alarm scent is strong enough, they fly into a frenzy and help attack the enemy. When only one ant discharges its scent glands, the effect is limited to a small area. When many release the pheromone, the scent may sweep through the nest, bringing out an army of furious defenders.

Other pheromones mark odor trails for workers to follow to sources of food or call large numbers of workers together when there is a job of work to be done, such as fixing a break in the nest's outer wall. A pheromone whose effects have been observed for centuries is released when a honeybee stings a human aggressor. If the attack takes place near the hive, other bees join the fray, drawn by a pheromone released when the first bee extrudes its sting.

One powerful pheromone is the queen substance, which is licked off the body of queen honeybees by their attendant workers and gets distributed around the nest. It serves as an inhibiting signal, preventing the workers from feeding and rearing larvae in a way that would make them develop into queens. If the reigning queen dies and the supply of queen substance is interrupted, its lack triggers chemical changes in the colony that permit the workers to develop young queens.

Pheromones also regulate the population of certain castes, especially among termites. Many species have a specialized military caste of soldiers armed with enormous jaws or with large glands full of poisonous or gluey liquids. There is evidence that this professional army is kept at proper strength by pheromones, some of which are released by the soldiers themselves. If one kind of pheromone predominates, fewer young soldiers appear. If another kind predominates, it is the signal for more to emerge.

As more pheromones are discovered every year, the subtlety of their uses becomes clearer. Most mammals are now known to employ them to some extent and some scientists think they may even influence human behavior. One expert, Edward O. Wilson of Harvard, has suggested: "It is conceivable that somewhere on other worlds civilizations exist that communicate entirely by the exchange of chemical substances that are smelled or tasted. . . . It is not difficult to design, on paper at least, a chemical communication system that can transmit a large amount of information with rather good efficiency."

The insects' chemical systems of communication and cooperation have never reached the point that Professor Wilson envisioned, but nevertheless insect society has proved so effective it is easy to overemphasize the similarity of insect and human communities. The differences are numerous and basic. The citizens of human nations are not predominantly sterile females that are descendants of a single egg-laying queen. When times are hard, human beings do not as a rule eat their babies. Neither does their system of communication depend on squirting perfumes out of glands that open all over their bodies.

Nevertheless, there is an important parallel between insect and human societies. Both were so successful they enabled their creators to move quickly into a great variety of ecological niches. Indeed, both social insects and social men founded rich ways of life that could not have been established except by organized groups, and some of these ways of life show astonishing similarities.

Two groups of tropical ants are nomadic hunters that sweep through the forests like conquering hordes, marching in columns to attack every living creature that cannot run or fly away. These driver and army ants can sometimes kill large reptiles, especially if they have been immobilized by a heavy meal. Human dwellers evacuate their villages to make way for the ants; when man returns, he finds his homes free of insects, spiders, centipedes, scorpions and lizards. Such efficient predation is perhaps not unexpected, but other insects engage in what seem to be almost civilized pursuits.

A very common and successful type of ant is reminiscent of human herdsmen who live principally on the milk of their cattle. The cattle of the ants are aphids (plant lice) or other small insects that suck the sap of plants. Aphids are common on tender shoots and other soft plant parts—in many cases because their ant proprietors have put them there to graze. Often the ants dig tunnels to take the aphids to places they could not reach on their own, such as plant roots. But even aboveground, ants can be seen guarding their small cows and milking them—stroking their backs with their antennae to make them exude a sweet fluid called honeydew, which the ants carry back to their nestmates.

Ants have never developed real agriculture to match their animal husbandry. It would be easy for the harvester ants, which store wild seeds for later consumption, to plant edible seeds and keep the seedlings clear of weeds, but apparently they have never taken advantage of the fact that planted seed will produce many more seeds of the same kind. However, a large group of the New World ant species known as leaf-cutters engages in a more unusual kind of farming: They grow fungus gardens in underground chambers following practices similar in many ways to commercial mushroom culture. These ants cut green leaves into handy fragments, carry them to underground chambers sometimes 20 feet deep and there chew and moisten them to a pulpy mass. They fertilize this culture medium with their feces and add bits of their special fungus that in many cases is not found except in ant nests. Weed types of fungus that may develop are removed. Soon the pulp is covered with fuzzy whitish threads and little round balls of fungus, the principal food of the ants.

The ants' fungus gardens permit them to make usable food of cellulose, the structural material of plants, which is very plentiful but is indigestible to animals above the level of the protozoans. The ants' fungus digests it and turns it into tender food on which the ants flourish. This scheme for upgrading an ordinarily unusable food source has been improved on only slightly by man, who cannot digest cellulose any better than ants can. Man feeds cellulose, in the form of hay, to cattle, which cannot digest it either but harbor stomach microorganisms that can. The cattle and their microorganisms help to convert the cellulose to protein—milk and meat—for human food. To the leaf-cutting ants the fungus is so vital that every young queen, when she flies off to begin a brand-new colony, will carry a pellet of it in a special mouth pouch.

It is hard to understand how tiny ant brains, aided by information-carrying chemicals, can do anything as complicated as fungus culture, which seems to require not only skill and knowledge but a good deal of foresight. Nevertheless they do it, and anyone who walks in a tropical American forest can see columns of leaf-cutting ants carrying angular bits of leaf toward a nest entrance. They march along roads cleared of obstacles. If leaf litter is dumped on a road, a repair squad quickly appears to clear it away. Human agriculturalists in the tropics hate the leaf-cutting ants, which can defoliate several cherished fruit trees in a single night, but they can do little against the ants. In some areas the ants make human agriculture almost impossible. They have been called "the real conquerors of Brazil."

Forms of agriculture and herding are not the only similarities to human society discernible among the ants, for the ant societies are amazingly diverse. Some kinds of ant are quiet and law-abiding, asking no more than to be left in peace to gather their seeds, milk their insect cows or cultivate their fungus gardens. Others are sneak thieves that dig slender tunnels among those of larger ants and emerge through small doorways to steal what they can. More violent ants raid the colonies of other ants, kill their workers and devour their helpless young. Some of these carry the young of a raided nest back to their own brood chambers. When the young captives emerge, they become slaves to their fierce masters.

A great many ant nests also harbor as guests other insects that are tolerated because they have in some fashion broken the code of their hosts' communication system. That is, they look like the host ants, mimic their movements, caress them in ways that please them and often secrete odorous substances attractive to them. There are thousands of varieties of these guest insects, including crickets, cockroaches, flies, mites, sow bugs and beetles. The guests pass as members of the colony and are fed as kinfolk instead of being killed as enemies—and, like permanent guests in human families, they may take over. Often the guests lay eggs that their hosts care for; when the alien eggs hatch and the larvae grow large, they eat the host ants' larvae.

One kind of ant welcomes certain guest-beetles that are even more diabolical in taking advantage of their hosts. The beetles' glands produce a substance to which the ants become so addicted that the whole colony is disorganized. If the nest is disturbed, the ants try first to save the drug-pushing beetles, carrying them to safety ahead of their own kind. The beetles breed in the ant colonies, where the ants feed and tend their larvae at the expense of their own, which often develop aberrantly. The beetle larvae reciprocate by eating the ants' young. They might well destroy the whole colony except for the saving fact that the ants' way of tending the beetle larvae when they are ready to pupate—covering them with soil and then uncovering them later—is fine for ants but death for beetles.

For some 50 to 100 million years such social insects have been one of the earth's dominant forms of

Wordless Languages of Love

Prancing, parading, puffing—each of the creatures on these pages is engaged in courtship. Courting is a serious business for all species because it is the prelude to mating—which, of course, is fundamental to survival. Feeling a sexual urge and lacking man's speech, other animals rely on wordless signals to recognize members of their own species and to communicate their desire. The male peacock spreads his gaudy feathers; a female howler monkey secretes a scent to entice males. Sexual communication can be as complex as the ballet of the female ostriches at bottom right or as simple as the croak of the male toad below, to which any nearby female toad will respond.

A toad's love call swells his vocal pouch to an enormous size.

To attract a mate, a male fiddler crab waves his claw, and an albatross stretches his wings, which may span almost 11 feet.

Competing for a lone male, female ostriches high-step on a Kenya savanna. When one female left the group, the male pursued her.

life. Though not as conspicuous as the large reptiles or mammals, their swarming numbers have always made up for their small size. They still do. In most parts of the modern earth the biomass (total body weight) and energy consumption of the ants alone exceed that of the vertebrates living in the same area.

The immensely successful and complicated world of the social insects is the product of association among individuals that otherwise could not have exploited their environments nearly as effectively. Then why shouldn't a similar device work for other animals? The great reptiles may have started to develop social organization—there is now some evidence that dinosaurs lived in groups (*Chapter 3*)—but they died out suddenly before they got very far. It remained for the mammals to surpass the insects and create the most far-reaching and effective social organizations the world has known.

The ultimate success of mammalian societies is something of a paradox, for mammals are physically constructed in ways that might seem to make group cooperation less useful for them than for insects.

On the face of it, all the pressures should have favored the individual over the group: Mammals may not necessarily need the competitive edge of society because they can grow much larger than insects without encountering serious difficulties. Their internal skeletons grow at the same pace as their bodies; they do not have to be shed and regenerated at each stage of growth—a costly and dangerous process for insects. And mammalian respiration, which uses lungs, a vigorous heart pump and a circulating blood system to carry oxygen to those tissues that need it, is a great deal more efficient for sizable animals than the insect system of fine tubes that pipe air directly to the oxygen-using tissues.

Most individual mammals are bigger than the total size of all the members of an ant, bee, wasp or termite colony. The mammals' size enables them to exploit more effectively many ecological opportunities, and it gives them powerful protection against most enemies, providing a degree of security that social insects attained by joint fighting action or by building strongly defended habitations.

Large size also makes mammals less vulnerable to cold, so they do not need group living to keep warm. Their big, insulated bodies and warm blood protect them against low temperatures that might kill an individual small insect or reduce it to dormancy. To avoid cold weather, ants, for example, may dig cooperative burrows deep into the earth, something that individual insects of similar size cannot do, at least not so effectively. But an individual mammal can defy the cold on the surface or easily dig its own snug burrow. Mammalian temperature control can also compensate for hot weather. So the joint action of many individuals in constructing weatherproof nests is not as necessary and does not encourage the formation of mammalian societies.

Mammals are also able to rear their young in ways that do not require the social organization on which insects depend. An important advantage gained by insects when they became social was the care that the workers of the colony could give the young of later generations. The insect nurses, protected in guarded chambers and supplied with food by foragers, protect, feed and clean the young until they reach full adulthood. Individual nonsocial insects cannot rear their offspring so safely and effectively, but individual mammals do something almost as good. Their young grow fairly large within their mothers' bodies and then are fed with milk and other food until they are able to take care of themselves. Here again, cooperation by many individuals is not as necessary and does not offer as great a premium for the formation of societies.

Despite all this, many mammals *did* form societies. The reason is that group living did yield them certain advantages. When mammals must live among predators, group living offers defensive advantages. Musk oxen protect themselves from circling wolves by forming a rough ring, horns outward, with the young in the center. The cries of prairie dogs spread through their towns the news of approaching danger. On the other hand, the predators may also benefit from cooperation. Modern wolves and wild dogs hunt in packs that can attack large prey more successfully and more safely than a single animal could, and it seems likely that their ancestors began to develop such cooperative behavior several million years ago. Similar social patterns that are now visible in other modern mammals must trace their beginnings to approximately the same time.

Groups such as these, if not too big, are usually made up of blood relatives. A social unit will normally include parents, children, the parents' brothers and sisters and their children, and perhaps an occasional "in-law." Sometimes the organization is very loose, but often the group has a recognized leader or several dominant adults, usually males, that will take precedence over the younger, weaker or less experienced members.

Such family societies are relatively crude. Sophisticated social groups among the mammals had to wait until the primates, man's ancestors, acquired fairly good brains. Some monkeys are so tightly organized into social groups that they appear unable to live singly. Baboons are elaborately social; their survival on the African veld is largely due to united action by the fighting males. Chimpanzees and gorillas have still other kinds of social organization. But neither approaches the level of even the most primitive forms of human society.

It was not until the appearance some five million years ago of man's immediate predecessor, the man-ape Australopithecus, that the development of complex social behavior began to accelerate. Only with a degree of social organization unprecedented among mammals could this four-and-a-half-foot-tall, 80-pound creature survive and prosper in a world dominated by bigger and more dangerous animals.

Australopithecines must have been loyal to their band, so the members could depend on each other in defense, hunting and food sharing. They probably divided up their duties, the females taking care of the young and gathering vegetable food and perhaps catching small game, while the males sought larger prey and looked out for defense. Since they were bipedal, they were capable of making and carrying primitive weapons such as wooden clubs. They must have had the wits both to use their weapons effectively and to outthink their prey. The hunting efficiency of these man-apes was a long time in developing, requiring many millions of years of experimentation, but although it depended on social organization it also furthered such organization.

Perhaps the biggest benefit derived from a successful hunting way of life was the encouragement it gave to the brain's development, for with the increasing complexity of the brain came that most human of achievements: speech. Just how or when human language was developed no one can say for sure, but it is fairly certain that the successor of Australopithecus, Homo erectus, the first true man, was on the way to developing a crude language.

Communication is the critical necessity of an advanced society, and speech is the best means of communication. The chemical pheromones employed so effectively by the social insects are generally useful only at short range; the visual signals of many animals work only when sender and receiver can see each other; the sounds that other animals use are limited mainly to warning signals and mating calls. Even with rudimentary speech early man could probably give commands ("You go straight ahead, while I circle around the hill"), transmit information ("I saw a lion across the ridge"), discuss plans ("Let's move camp") and, above all, diffuse information throughout the band. With language, the experience of each member became the experience of all. Even the knowledge of members long dead was still useful; it contributed to the band's traditions and accumulated culture. Equipped with a simple but sufficient language, Homo erectus was firmly on the road to modern man, although his enormous future success was not yet apparent.

If outside observers had toured the earth half a million years ago, they probably would not have selected Erectus as particularly important. He was not numerous, and in comparison with the planet's larger animals he was physically weak, as modern man is. But he had a fairly large and growing brain, rudimentary speech and a constantly improving culture that he could pass on from generation to generation. These are the essentials that enabled his modern descendant, Homo sapiens, to create the superorganisms of human societies and conquer the earth.

Since the appearance of Sapiens about 300,000 years ago, man's biological evolution has been overshadowed by his cultural and social evolution. Physically he has not changed much. No Stone Age man, properly clothed and barbered, would attract attention on the streets of New York or Paris. What has made modern man modern is the success of his spectacular and rapidly evolving society.

Mirrors of Man's Society

Driver ants march through an African rain forest in a disciplined column, soldiers with raised pincers flanking the mass of workers.

Twenty million strong, a whole colony of driver ants moves out in a well-shaped column (*above*) to build a new nest, after exhausting the resources within reach of their old one. They may march for three days and as far as a mile—the equivalent of 125 miles for a man. And at every moment their long trek will depend upon complex social behavior, in which the actions of each insect contribute to the successful completion of a common task.

In this ancient society—as in many other social organizations among animals—man finds counterparts to his own behavior, for all societies depend on a division of labor and cooperation toward common objectives. These counterparts to human life are isolated, rather than combined in the uniquely human way, but they are there. Thus prairie dogs talk to one another, using a rudimentary yet precise language of vocal calls. Titi monkeys apparently pair off in lifelong relationships. The pampered young of chimpanzees are taught the work and ways of adult life. But not until the evolution of the hunting bands of Homo erectus, more than half a million years ago, did all three elements—language, love and learning—come together in the powerful society of man.

In a termite mound, worker termites, their abdomens dark with a nutritious mixture of mud and humus, feed a huge queen-mother.

The Automated Life of a Termite Mound

In special nursery chambers, eggs hatch and adult workers tend the growing infants.

Workers rush to patch a breach in the mound wall, using mud cemented with saliva.

In mounds as complex as many a human city, termites duplicate some of prehistory's earliest organized patterns of social behavior—patterns that may have been laid down more than 100 million years ago, when dinosaurs walked the earth.

A typical mound, like that of the African genus *Cubitermes (left)*, contains a king and queen and thousands of their progeny. This huge colony works together by a caste system more rigid than any tried by man.

Normally, only the king and queen are fertile, and they do nothing but breed; their sterile offspring perform all other tasks. Soldiers, about 3 per cent of the population, defend the mound against such enemies as ants. The workers maintain the mound, groom the king and queen and care for the young. They even eat for everybody, doling out predigested food from their own abdomens.

This complex behavior is controlled by chemical signals. The termites are blind and deaf; their senses of smell and taste, responsive to secretions called pheromones, serve for communication. Thus, new-laid eggs issue a chemical command to the workers that says, in effect: "Take us to a hatchery!" Even the size of the population is regulated by pheromones; for example, when the mound contains too many soldiers, the excess of soldier pheromone prompts workers to attack them and devour the surplus.

A Code of Yips and Yelps in a Prairie Dog Town

Sounding an exuberant two-note call, a prairie dog pup leaps up to signal "all's well."

An adult dog alerts the entire township to danger with a series of short, nasal yips.

The rodents called prairie dogs have developed a type of society so successful that hundreds of millions of these animals flourished in the Great Plains before the white man came.

A major element in their success is the vocabulary of barks and calls that earned the rodents their misleading name—a language that is one of the most precise systems of communication in nonhuman societies. With its help, the inhabitants of a prairie dog town, many acres or even square miles in extent, gain superb protection against a wide range of natural enemies—eagles, hawks, coyotes, bobcats and badgers.

A town is divided into independent territories, each inhabited by a clan, or coterie, with an average membership of one male, three females and about six pups. The members of a coterie rarely stray outside the boundaries of their territory, and they fiercely resist all intruders. But they and all their neighbors are bound together by the language of prairie dog calls. A single sharp call across the flatland alerts a grazing prairie dog to sit up and peer about for danger. A series of short, high barks is a full-scale alarm that sends all the creatures scurrying underground. Finally, when danger has passed, a single animal balances on its hind legs to sound the two-note "all clear" call—and the whole town resounds as the jubilant signal is taken up from burrow to burrow.

A prairie dog family takes the air outside a burrow. The raised entrance is both a vantage point and a barrier against flash floods.

Among the Titis an Exclusive Society for Two

The titi monkeys of South America are believed to pair off for life. This practice is habitually followed by only two other primate species, the gibbon and, most of the time, man. In man pairing off normally involves the complex institution called marriage and the practice of sexual fidelity. The titis' main interest is companionship; sexually they are promiscuous. The warmth of their lifelong relationship is expressed in nuzzling, in mutual grooming and in a snuggling perch with entwined tails, the posture in which they habitually sleep. But once a year, at the breeding season, they separate for short-lived sexual encounters with neighbors of the opposite sex. Afterward, the pair calmly takes up the threads of normal life, untroubled by jealousy, and the male apparently cherishes its mate's young without regard to their paternity.

This apparently warm relationship within the titi family is counterbalanced by ferocious territorial jealousy between families. Every morning the couple advances to the boundary of its acre-plus territory and screeches defiance at its neighbors; the screeching becomes steadily louder, and the couple's backs arch, their hair bristles and their tails lash in fury. If a neighbor trespasses even in the most minor way, the aggressive display may lead to a chase and a sudden nip. Such antagonism between families may relieve tensions between individuals.

Tails companionably entwined, two titi monkeys eye the world with calm assurance.

A Titi stretches out luxuriously while its mate combs its fluffy coat. Titis often spend much of their day in such mutual grooming.

The Free and Flexible Life of the Apes

Of all animal societies the closest to man's—in its stress on personal relations, individual learning and flexibility—is that of the chimpanzee. Few societies are more loosely structured; group composition changes constantly. Within a group, the strongest males generally dominate, but the hierarchy is not rigid. A small but daring chimp can win leadership in a boast-and-swagger duel that ends in a spell of mutual grooming.

Perhaps most humanlike of all the chimp's social characteristics is the intimate relation between mother and child. A baby suckles to the age of four and does not reach puberty until it is about eight. During these years of personal development, it learns to use rudimentary tools: branches for nest building, sticks and stones for weapons, chewed leaves for sponges. And it accompanies these activities with a never-ending stream of communicative gestures, grunts, calls, hoots, barks and squeals—along with a rich vocabulary of facial expressions.

A frisky chimpanzee baby reaches for attention from a friendly male. He will not be rejected; adults love to fondle the young.

The Organized Hunting Band of the First True Men

The picture at right combines photography and painting to show how the first true humans lived some 600,000 years ago. This organized band of the now-extinct species *Homo erectus* has journeyed to a dry river bed in search of quartz and chert—rocks that they can chip into tools and weapons but can find easily only here. Their mission is dangerous. Icy winds sweep this Asian plain, and predators such as saber-toothed cats are on the prowl. The band may have invaded another group's territory, and to the "owners" of the valuable quarry the invaders would be game as fair—and as tasty—as any other.

If the daring expedition succeeds, it will do so largely because its members work together well, in a complex yet flexible social group. Tasks are divided up: some men forage for fuel and hunt game; others, perhaps the world's first skilled craftsmen, make tools. A woman tends the fire, nurses a baby and keeps an eye on a youngster. And to help them attain their objective, they have assets that are unique to man. Even these early humans use a rudimentary language to make plans and exchange information. They have good weapons, like the all-purpose chopper brandished by the man at far right. They have fire, transported as glowing embers in hide sacks. Forever improving their skills and organization, bands like this one in time came to dominate the earth.

Weapons at the ready, a skin-clad Homo erectus hunter glares out at a hostile world as he rejoins his band at a temporary camp.

The Emergence of Man

This chart records the progression of life on earth from its first appearance in the waters of the new-formed planet through the evolution of man; it traces his physical, social, technological and intellectual development to the Christian era. To place these advances in commonly used chronological sequences, the column

Geology	Archeology	Billions of Years Ago	
Precambrian earliest era		4.5	Creation of the Earth
		4	Formation of the primordial sea
			First life, single-celled algae and bacteria, appears in water
		3	
		2	
		1	
		Millions of Years Ago	
			First oxygen-breathing animals appear
		800	
			Primitive organisms develop interdependent specialized cells
		600	Shell-bearing multicelled invertebrate animals
Paleozoic ancient life			Evolution of armored fish, first animals to possess backbones
		400	Small amphibians venture onto land
			Reptiles and insects arise
			Thecodont, ancestor of dinosaurs, arises
Mesozoic middle life		200	Age of Dinosaurs begins
			Birds appear
			Mammals live in shadow of dinosaurs
			Age of Dinosaurs ends
		80	
Cenozoic recent life			Prosimians, earliest primates, develop in trees
		60	
		40	Monkeys and apes evolve
		20	
		10	Ramapithecus, oldest known primate with apparently manlike traits, evolves in India and Africa
		8	
		6	
		4	Australopithecus, closest primate ancestor to man, appears in Africa

Geology	Archeology	Millions of Years Ago	
Lower Pleistocene oldest period of most recent epoch	**Lower Paleolithic** oldest period of Old Stone Age	2	Oldest known tool fashioned by man in Africa
			First true man, Homo erectus, emerges in East Indies and Africa
		1	Homo erectus migrates throughout Old World tropics
		Thousands of Years Ago	
Middle Pleistocene middle period of most recent epoch		800	Homo erectus populates temperate zones
			Man learns to control and use fire
		600	
			Large-scale, organized elephant hunts staged in Europe
		400	Man begins to make artificial shelters from branches
		200	
Upper Pleistocene latest period of most recent epoch	**Middle Paleolithic** middle period of Old Stone Age		Neanderthal man emerges in Europe
		80	
Last Ice Age		60	Ritual burials in Europe and Near East suggest belief in afterlife
			Woolly mammoths hunted by Neanderthals in northern Europe
		40	Cave bear becomes focus of cult in Europe
	Upper Paleolithic latest period of Old Stone Age	30	Cro-Magnon man arises in Europe
			Man reaches Australia
			Oldest known written record, a lunar calendar on bone, made in Europe
			Asian hunters cross Bering Strait to populate North and South America
			Figurines sculpted for nature worship
			First artists decorate walls and ceilings of caves in France and Spain
		20	Invention of needle makes sewing possible
			Bison hunting begins on Great Plains of North America
Holocene present epoch	**Mesolithic** Middle Stone Age	10	Bow and arrow invented in Europe
			Dog domesticated in North America

Four billion years ago
Three billion years ago
Origin of the Earth (4.5 billion)
First life (3.5 billion)

at the far left of each of the chart's four sections identifies the great geological eras into which earth history is divided, while the second column lists the archeological ages of human history. The key dates in the rise of life and of man's outstanding accomplishments appear in the third column (years and events mentioned in this volume of The Emergence of Man appear in bold type). The chart is not to scale; the reason is made clear by the bar below, which represents in linear scale the 4.5 billion years spanned by the chart—on the scaled bar, the portion relating to the total period of known human existence (far right) is too small to be distinguished.

Geology	Archeology	Years B.C.	
Holocene *(cont.)*	**Mesolithic** *(cont.)*	9000	Jericho settled as the first town
			Sheep domesticated in Near East
	Neolithic New Stone Age	8000	Pottery first made in Japan
			Goat domesticated in Persia
			Man cultivates his first crops, wheat and barley, in Near East
		7000	Pattern of village life grows in Near East
			Catal Huyuk, in what is now Turkey, becomes the first trading center
			Loom invented in Near East
			Agriculture begins to replace hunting in Europe
		6000	Cattle domesticated in Near East
			Copper used in trade in Mediterranean area
	Copper Age		Corn cultivated in Mexico
		4000	Sail-propelled boats used in Egypt
			Oldest known massive stone monument built in Brittany
			First cities rise on plains of Sumer
			Cylinder seals begin to be used as marks of identification in Near East
		3500	First potatoes grown in South America
			Wheel originates in Sumer
			Egyptian merchant trading ships start to ply the Mediterranean
			First writing, pictographic, composed, Near East
	Bronze Age	3000	Bronze first used to make tools in Near East
			City life spreads to Nile Valley
			Plow is developed in Near East
			Accurate calendar based on stellar observation devised in Egypt
			Sumerians invent potter's wheel
			Silk moth domesticated in China
			Minoan navigators begin to venture into seas beyond the Mediterranean
		2600	Variety of gods and heroes glorified in *Gilgamesh* and other epics in Near East
			Pyramids built in Egypt
		2500	Cities rise in the Indus Valley

Geology	Archeology	Years B.C.	
Holocene *(cont.)*	**Bronze Age** *(cont.)*	2400	Stonehenge, most famous of ancient stone monuments, begun in England
			Earliest written code of laws drawn up in Sumer
		2000	Chicken and elephant domesticated in Indus Valley
			Use of bronze spreads to Europe
			Eskimo culture begins in Bering Strait area
			Man begins to cultivate rice in Far East
			Herdsmen of Central Asia learn to tame and ride horses
		1500	Invention of ocean-going outrigger canoes enables man to reach islands of South Pacific
			Oldest known paved roads built in Crete
			Ceremonial bronze sculptures created in China
			Imperial government, ruling distant provinces, established by Hittites
	Iron Age	1400	Iron in use in Near East
			First complete alphabet devised in script of the Ugarit people in Syria
			Hebrews introduce concept of monotheism
		1000	Reindeer domesticated in northern Europe
		900	Phoenicians develop modern alphabet
		800	Celtic culture begins to spread use of iron throughout Europe
			Nomads create a far-flung society based on the horse in Russian steppes
			First highway system built in Assyria
			Homer composes *Iliad* and *Odyssey*
		700	Rome founded
			Wheel barrow invented in China
		200	Epics about India's gods and heroes, the *Mahabharata* and *Ramayana*, written
			Water wheel invented in Near East
		0	Christian era begins

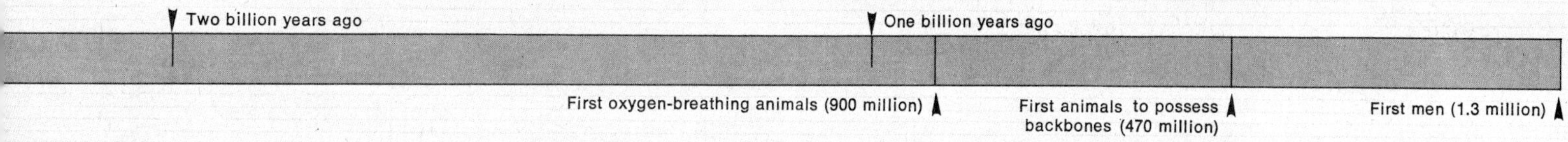

Credits

The sources for the illustrations in this book are shown below. Credits from left to right are separated by semicolons, from top to bottom by dashes.

Cover—Painting by Burt Silverman, background photograph by Otto Lang from Photophile. 8—Harald Sund. 12, 13—Drawings by Robert McKee, after photographs by E. Muybridge courtesy Dover Publications, Inc. 16, 17—Ken Kay. 20, 21—Nina Leen for LIFE, drawings by Adolph E. Brotman. 24 through 47—Paintings by Don Punchatz. 49 —Fritz Goro, Peabody Museum of Natural History, Yale University. 50—Fritz Goro courtesy Dr. T. Delevoryas—Dr. Frank M. Carpenter. 51—Fritz Goro, Princeton University Museum of Natural History. 52—Henry B. Beville, Smithsonian Institution. 53—Fritz Goro, Museum of Comparative Zoology, Harvard University—Dr. Frank M. Carpenter. 54, 55—Henry B. Beville, Smithsonian Institution. 56—Fritz Goro courtesy of The American Museum of Natural History. 57—Fritz Goro, Museum of Comparative Zoology, Harvard University—Fritz Goro, Peabody Museum of Natural History, Yale University. 58 —Fritz Goro, specimen in the British Museum (Natural History)—Fritz Goro, Peabody Museum of Natural History, Yale University. 59—Henry B. Beville, Smithsonian Institution—Fritz Goro, Peabody Museum of Natural History, Yale University. 60 —Fritz Goro, Museum of Comparative Zoology, Harvard University. 61—Fritz Goro, Botanical Museum, Harvard University. 62 —Willard Starks, Princeton University Museum of Natural History. 64—Drawings by Adolph E. Brotman, based upon book *Men and Dinosaurs* by Edwin H. Colbert. Copyright © 1968 by Edwin H. Colbert. Published by E. P. Dutton & Co., and used with their permission. 66 through 71—Paper sculpture by Nicholas Fasciano, photographed by Ken Kay. 74—John R. Freeman courtesy The Royal Society, London; Culver Pictures. 75, 76, 77—The Mansell Collection, London. 78 —Courtesy of The American Museum of Natural History. 79—Peabody Museum of Natural History, Yale University. 81—Watercolors by Arthur Lakes, photographed by Fritz Goro; Paulus Leeser (2), Peabody Museum of Natural History, Yale University. 82 —Watercolors by Arthur Lakes, photographed by Fritz Goro—Paulus Leeser (2), Peabody Museum of Natural History, Yale University. 83—Watercolors by Arthur Lakes, photographed by Benschneider courtesy of the Arthur Lakes Library of the Colorado School of Mines, Golden. 86, 87 —The Palaeozoological Institute of the Polish Academy of Sciences in Warsaw, courtesy of Dr. Zofia Kielan-Jaworowska, photograph by W. Skarżyński. 89 through 97 —Paintings by Burt Silverman. 98—Lennart Nilsson. 102 through 111—Drawings by Nicholas Fasciano. 117 through 121—Drawings by Nicholas Fasciano. 120, 121—Opossum, Rhesus, Chimpanzee musculature based on figures in Huber, 1930 (*Quart. Rev. Biol.*, 5: 147, 404, 408), and redrawn with the permission of the editors of *The Quarterly Review of Biology*. 122—John Dominis for LIFE. 126—Dr. Roman Vishniac. 134—David Hughes from Bruce Coleman, Inc. 135—Shelly Grossman; Niall Rankin—Akhtar Hussein. 139—Carlo Bavagnoli for LIFE. 140, 141—Dr. Edward S. Ross. 142—Arthur Rickerby for LIFE. 143 —Leonard Lee Rue III from FPG. 144, 145—Nina Leen for LIFE. 146, 147— Dr. Phyllis Dolhinow courtesy Gombe Stream Reserve Research Centre. 148, 149 —Painting by Burt Silverman, background photograph by Dean Brown.

Acknowledgments

For the help given in the preparation of this book, the editors are indebted to Stanley Awramik, Museum of Comparative Zoology, Harvard University; Donald Baird, Curator, Department of Geological and Geophysical Sciences, Princeton University; Umesh Banerjee, Department of Geological Sciences, Harvard University; Elso S. Barghoorn, Professor of Botany, Harvard University; Joan Bliss, Office of the Councillor Scientific, Australian Embassy, Washington, D.C.; Richard Bremner, Programming Manager for Illiac, Burroughs Corporation, Paoli, Pa.; Carol Campbell, Museum of Comparative Zoology, Harvard University; Frank Carpenter, Fisher Professor of Natural History, Harvard University; Robert L. Carroll, Redpath Museum, McGill University, Montreal; Joseph Cope, Gilbert Cope Foundation, West Chester, Pa.; Ross Dalen, Control Data Corporation, Minneapolis; Irven DeVore, Professor of Anthropology, Department of Social Relations, Harvard University; Peter Dodson, Department of Geology and Geophysics, Peabody Museum of Natural History, Yale University; Daniel Fisher, Museum of Comparative Zoology, Harvard University; Roger Sheridan Fouts, Department of Psychology, University of Oklahoma, Norman; Richard D. Fox, Associate Professor of Geology and Zoology, University of Alberta, Edmonton; Peter Galton, Department of Zoology, University of Bridgeport, Conn.; Shirley Hartman, Chief of Exhibit Design, Peabody Museum of Natural History, Yale University; Nicholas Hotton III, Curator, Fossil Amphibians and Reptiles, Department of Paleobiology, National Museum of Natural History, Smithsonian Institution, Washington, D.C.; James A. Jensen, Curator, Earth Sciences Museum, Brigham Young University, Dinosaur National Museum, Provo, Utah; Glenn L. Jepsen, Professor of Vertebrate Paleontology, Princeton University; Richard Kay, Peabody Museum of Natural History, Yale University; Zofia Kielan-Jaworowska, Polska Akademia Nauk, Warsaw, Poland; Herbert Killackey, Visiting Assistant Professor, Division of Biological Sciences, Brown University, Providence, R.I.; Victoria Kohler, Museum of Comparative Zoology, Harvard University; Frank Long, Film Officer, Australian News and Information Bureau, New York City; Copeland McClintock, Assistant to the Director, Peabody Museum of Natural History, Yale University; Jesse Merrida, Museum Specialist, Department of Paleobiology, National Museum of Natural History, Smithsonian Institution, Washington, D.C.; Judith H. Metzger, Research Assistant, Department of Geological Sciences, Harvard University; Lorus Milne, Professor of Zoology, University of New Hampshire, Durham; David R. Pilbeam, Associate Professor of Anthropology, Yale University; Edward S. Ross, California Academy of Sciences, Golden Gate Park, San

Francisco; Dale Russell, Chief, Paleontology Division, National Museum of Natural Sciences of Canada, Ottawa; Bobb Schaeffer, Chairman and Curator, Department of Vertebrate Paleontology, American Museum of Natural History, New York City; Charles Schaff, Museum of Comparative Zoology, Harvard University; Charles Smart, Department of Geological and Geophysical Sciences, Princeton University; William N. Tavolga, Research Associate, Department of Animal Behavior, American Museum of Natural History, New York City; C. Richard Taylor, Director, Concord Field Station, Bedford, Mass.; Richard H. Tedford, Curator, Department of Vertebrate Paleontology, American Museum of Natural History, New York City; Keith S. Thomson, Associate Professor of Biology, Peabody Museum, Yale University; Howard Topoff, Research Associate, American Museum of Natural History, New York City; Karl M. Waage, Professor, Department of Geology and Geophysics, Peabody Museum of Natural History, Yale University.

Bibliography

Geology and Paleobotany

Andrews, H. N., Jr., *Ancient Plants and the World They Lived In.* Comstock Publishing, 1947.

Arnold, C. A., *An Introduction to Paleobotany.* McGraw-Hill, 1947.

Brooks, C. E., *Climate Through the Ages.* Dover Publications, 1971.

Croneis, Carey G., and William C. Krunbein, *Down to Earth: An Introduction to Geology.* University of Chicago Press, 1971.

Delevoryas, Theodore:
Plant Diversification. Holt, Rinehart and Winston, 1966.
Morphology and Evolution of Fossil Plants. Holt, Rinehart and Winston, 1962.

Dunbar, Carl, and Karl Waage, *Historical Geology.* John Wiley & Sons, 1969.

Kummel, Bernard, *History of the Earth.* W. H. Freeman, 1970.

Leet, Lewis E., and Florence J. Leet, eds., *World of Geology.* McGraw-Hill, 1971.

Stokes, William, *Essentials of Earth History.* Prentice-Hall, 1966.

Tarling, Don, and Maureen Tarling, *Continental Drift: A Study of the Earth's Moving Surface.* Doubleday, 1971.

Biology and Evolution

Barnes, Robert D., *Invertebrate Zoology.* W. B. Saunders, 1968.

Colbert, Edwin, *Evolution of the Vertebrates.* John Wiley & Sons, 1969.

Gregory, W. K., *Our Face from Fish to Man.* Capricorn Books, 1965.

Hotton, Nicholas, III, *The Evidence of Evolution.* American Heritage, 1968.

Hyman, Libbie Henrietta, *The Invertebrates: Protozoa through Ctenophora,* Vol. 1. McGraw-Hill, 1940.

Le Gros Clark, Wilfred, *The Antecedents of Man.* Quadrangle Books, 1960.

McAlester, A. Lee, *The History of Life.* Prentice-Hall, 1968.

McCulloch, Gordon, *Man and His Body.* The Natural History Press, 1967.

Mercer, E. H., *Cells: Their Structure and Function.* Doubleday, 1967.

Moment, Gardner B., *General Zoology.* Houghton Mifflin, 1958.

Noble, G. Kingsley, *The Biology of the Amphibia.* Dover Publications, 1954.

Romer, Alfred Sherwood:
Vertebrate Paleontology. University of Chicago Press, 1966.
The Vertebrate Story. The University of Chicago Press, 1959.

Simpson, G. G., C. S. Pittendrigh, and L. H. Tiffany, *Life: An Introduction to Biology.* Harcourt, Brace, 1957.

Tobias, Philip V., ed., *The Functional and Evolutionary Biology of Primates.* Aldine Atherton, 1972.

Wells, Martin, *Lower Animals.* McGraw-Hill, 1968.

Dinosaurs

Colbert, Edwin:
Men and Dinosaurs. McGraw-Hill, 1951.
The Dinosaur Book. McGraw-Hill, 1951.
Dinosaurs, Their Discovery and Their World. E. P. Dutton, 1961.

Kurten, Bjorn, *The Age of the Dinosaurs.* McGraw-Hill, 1968.

Ostrom, John H., *The Strange World of Dinosaurs.* G. P. Putnam's Sons, 1964.

Ostrom, John H., and John S. McIntosh, *Marsh's Dinosaurs.* Yale University Press, 1966.

Swinton, W. E., *The Dinosaurs.* Thomas Murby, 1934.

Animal Behavior

Bellairs, Angus d'A., *Reptiles.* Harper Torchbooks, 1960.

Chance, Michael, and Clifford Jolly, *Social Groups of Monkeys, Apes and Men.* E. P. Dutton, 1970.

Dethier, V. G., and Eliot Stellar, *Animal Behavior.* Prentice-Hall, 1961.

DeVore, Irven, ed., *Primate Behavior.* Holt, Rinehart and Winston, 1965.

Goetsch, Wilhelm, *The Ants.* University of Michigan Press, 1969.

Gray, James, *How Animals Move.* Cambridge University Press, 1953.

Gregory, R. L., *Eye and Brain.* McGraw-Hill, 1966.

Howell, A. Brazier, *Speed in Animals.* University of Chicago Press, 1944.

Howse, P. E., *Termites.* Hutchinson University Library, 1970.

Hutchins, Ross E., *The Ant Realm.* Dodd, Mead & Company, 1967.

Jay, Phyllis C., ed., *Primates.* Holt, Rinehart and Winston, 1968.

Matthews, L. Harrison, and Maxwell Knight, *The Senses of Animals.* Museum Press, 1963.

Mech, L. Davis, *The Wolf.* The Natural History Press, 1970.

Michener, Charles D., and Mary H. Michener, *American Social Insects.* D. Van Nostrand, 1951.

Muybridge, Eadweard:
The Human Figure in Motion. Dover Publications, 1955.
Animals in Motion. Dover Publications, 1957.

Schiller, Claire H., *Instinctive Behavior.* International Universities Press, 1964.

Smythe, R. J., *Animal Vision: What Animals See.* Charles C. Thomas, 1961.

Tavolga, William N., *Principles of Animal Behavior.* Harper & Row, 1969.

Van Lawick-Goodall, Jane, *In the Shadow of Man.* Houghton Mifflin, 1971.

Wilson, Edward O., *The Insect Societies.* The Belknap Press of Harvard University Press, 1971.

Index

Numerals in italics indicate an illustration of the subject mentioned.

Printed in U.S.A.